PATENTING LIFE?
STOP!

SEÁN McDONAGH

Patenting Life?

Stop!

*Is Corporate Greed
Forcing Us to Eat
Genetically Engineered Food?*

DOMINICAN PUBLICATIONS

First published (2003) by
Dominican Publications
42 Parnell Square
Dublin 1

ISBN 1-871552-85-0

British Library Cataloguing in Publications Data.
A catalogue record for this book is available
from the British Library.

Copyright © (2003) Seán McDonagh

All rights reserved.
No part of this publication may be reproduced, stored in
a retrieval system or transmitted by any means,
electronic or mechanical, including photocopying,
without permission in writing from the publisher.

Cover design by Bill Bolger

Printed in Ireland by
The Leinster Leader Ltd
Naas, Co. Kildare.

For Eamon O'Hara,
always willing to lend a hand

Contents

Glossary	9
Introduction	11
1. A World Ruled by Transnational Corporations	15
2. From Agriculture to Agribusiness	44
3. The ABC of Genetic Engineering	69
4. The Pros and Cons of GE Food	78
5. An Unholy Trinity – Regulatory Agencies, Biotech Corporations, and Governments	115
6. Ethics and Genetic Engineering	142
7. Stop Patenting Life	175
Conclusion	221
Further Information	223

Glossary

BAT	British and American Tobacco
Bt	Bacillus thuriengensis
CBD	Convention on Biodiversity
CFCs	Chlorofluorocarbons
CORI	Conference of Religious of Ireland
CDC	Center for Disease Control
DAARE	Disabled Against Animal Research and Exploitation
DNA	Deoxyribonucleic acid
ERT	European Round Table of Industrialists
EU	European Union
EPA	Environmental Protection Agency
FAO	Food and Agricultural Organisation (UN)
FDA	Food and Drug Administration (US)
GATT	General Agreement on Tariffs and Trade
GATS	General Agreement on Trade and Services
GDP	Gross Domestic Product
IPPC	International Panel on Climate Change (UN)
OPIC	Overseas Private Investment Corporation (US)
MTAs	Material Transfer Agreements
NOAA	National Oceanic and Atmospheric Administration (US)
PCBs	Polychlorinated biphenyls
PVP	Plant Variety Protection
TBT	Technical Barriers to Trade
TRIPs	Trade Related Intellectual Properties
UNCED	United Nations Conference on Environment and Development
UNDP	United Nations Development Programme
UNICEF	United Nations Children's Fund

UPOV	Union for the Protection of New Varieties of Plants
WHO	World Health Organisation
WTO	World Trade Organisation
WWF	World Wildlife Fund

Introduction

I AM AN Irish Columban missionary priest. Immediately after ordination in 1969 I was assigned to the Republic of the Philippines where I spent the next twenty-one years working as a missionary on the island of Mindanao. During my time there, as part of my missionary ministry I became involved in a variety of programmes and campaigns centring on the themes of peace, justice and creation.

In 1972, the President of the Philippines, Ferdinand Marcos, declared martial law and effectively became a dictator. Many Church workers and others who had campaigned for human rights, land reform and other justice issues where arrested, tortured and killed. While promoting justice and human rights, the Catholic Church saved the lives of many people. It also mobilised 'people power' in 1986 to overthrow the Marcos regime.

Between 1980 and 1991, I ministered among a tribal people called the T'boli, in the hills of South Cotabato. It was there that I had first-hand experience of the devastation tropical deforestation wreaks on the environment and on people. I had what Pope John Paul II would call, almost two decades later, an 'ecological conversion'. I began to speak and, later, write about the desecration of God's creation in an effort to raise awareness of this crucial issue, especially among Church people.

My first book, *To Care for the Earth* (1985), was the first on ecology and theology published in Britain and Ireland. In the years since then I have written numerous articles and books on ecological and justice issues in the context of our Christian faith. These have covered topics like Third World debt, deforestation, global warming, soil erosion, incineration, nuclear power, population pressures, water pollution and the devastation of the oceans. I helped draft the Philippine Bishops' pastoral letter on the environment, *What is Happening to Our Beautiful Land?* This was the first pastoral letter from any episcopal conference exclusively on the environment.

In my 1999 book, *Greening the Christian Millennium*, one chapter, entitled 'Ethics and Genetic Engineering', looked at the use of genetic engineering in agriculture. Some might wonder why I choose to address this topic which might, at first glance, seem somewhat esoteric and of little interest for a Christian writer. My reason is simple: genetic engineering is about how food is produced and who controls this process. It is a vital contemporary ethical issue because humans, or, for that matter, any other creature cannot live without food.

This book, *Patenting Life? Stop!*, greatly expands on my previous presentation. It explores how transnational corporations have cornered agriculture and the food industry. The subtitle of the book – *Is Corporate Green Forcing Us to Eat Genetically Engineered Food?* – argues that these corporations are promoting biotechnology even though many independent scientists have genuine fears about the safety of foods so produced, for human consumption. Other scientists argue that GE agriculture will have a major negative impact on the environment and will undermine biodiversity which is already threatened by the current extinction spasm that has gripped the living world. If the corporations are successful, a handful of them will control the seeds of all the staple crops globally. This would be a nightmare for food security.

The final chapter contends that the rush to patent life will devalue all life and, instead of feeding the world, will exacerbate hunger and malnutrition. The claim to patent life is a break with all previous cultural and religious traditions and an affront to them. Seeing a trade agreement within the World Trade Organisation (WTO) used as a battering ram to promote this legislation globally gives us some glimpse of the bullying tactics which successive US administrations have used to advance the interests of US transnational corporations even when the lives of tens of millions of people are at stake.

I wish to thank Fr Bernard Treacy, O.P., for his patient work on this text during the past year. A word of thanks also to my colleagues Fr Oliver Kennedy, Fr Pat Connaughton, Fr Paddy Coneally, Fr Michael Duffy, Fr Noel Doyle and Fr Brendan MacHale for their many helpful suggestions as they proofread the text. And, finally, thanks to my sister, Máire, for all her support while I was writing this book.

ONE

A World Ruled by Transnational Corporations

TRANSNATIONAL CORPORATIONS (TNCs) are the most important economic and commercial organisations in the modern world. They have their roots in the Church and the State. The ecclesiastic roots go back to feudal times in Europe. The ruling class passed on their property to their children. Because the leadership of the Church and, especially of the monastic communities, was composed of celibate individuals, their property was held by the institution collectively. This gave rise to the idea that an institution could own property and be considered a moral person.

Secular corporations go back to state-sponsored corporations set up by colonial powers from the sixteenth century onwards. One of the most prominent of these was the British East India Company established under royal charter by Queen Elizabeth I, to trade in spices. Its capital was £70,000. From its foundation until its dissolution in 1858, it grew into a powerful trading, political and military entity. During the eighteenth and nineteenth centuries the Company promoted British interests throughout the Indian subcontinent, reaching as far as Burma. India supplied raw material for the growing industrial revolution in Britain but also became a dumping ground for cheap clothes and industrial goods that were being mass-produced in Britain. These economic policies wiped out local industry by imposing prohibitive duties on indigenous goods. This led to hardship for millions of Indians. The cruel management policies of the Company led to mutiny in 1857. One year later, after a parliamentary enquiry, the company was dissolved and its assets taken over by the British government.

On the other side of the Atlantic, land corporations, given charters by the crown, played an important part in the white settlement of North America. Both the Virginia Company and the New England Company were chartered in 1606. The building of canals and the railway marked another point in the evolution of corporations in North America in the 1820s, 1830s and 1840s.

The transnational phase of modern corporations began to expand dramatically after the Civil War in the United States in the 1860s. The resources of the world – oil, coal, ore, land and timber – were there to be exploited and transported to places where they could be used to make profit. After the 1880s, with the advent of the new electrical, mechanical and chemical technologies, corporations used scientific research to develop a whole variety of products for industry and households. Selling these products gave rise to the advertising industry. Some of the industries that were founded at that time – Standard Oil Company of Ohio (1870), General Electric (1892), Dow Chemical (1897), Johnson & Johnson (1887), Carnegie Steel (1873) and Ford Motors (1903) – are still household names in the US today.

Finally, at the dawning of this new corporate era, a firm and corporate-friendly legal framework was already in place in the US. Corporate goals took precedence over everything else in the legal system. As Morton Horowitz, chair of the History of Law at Harvard University, writes in his book, *The Transformation of American Law: 1780-1860*: 'By the middle of the nineteenth century the legal system had been reshaped to the advantage of men of commerce and industry at the expense of farmers, workers, consumers, and other less powerful groups within society.'[1] This allowed the nascent corporations to mobilise capital on an unprecedented scale, opening wide the floodgates for horizontal and vertical integration of cor-

---

1. Quoted in Thomas Berry, *The Great Work: Our Way Into The Future*, Bell Tower, New York, p. 143.

porate structures. The national importance of the corporations was summed up in the slogan, 'What is good for General Motors is good for America.'

The crucial element in these newly established trading entities was the notion of limited liability. A shareholder could not be held liable for an amount greater than his or her investment. The purpose of this was to encourage entrepreneurs to take risks. Many commentators at the time were opposed to the limited liability. They argued that it would lead to excessive risk-taking by managers and owners or what is today called 'moral hazard'. Adam Smith, for example, held this view. According to the economist Ha-Joon Chang, 'limited liability provides one of the most powerful mechanisms to "socialize risk" which has made possible investments of unprecedented scale'.[2]

Much of the expansion of TNCs in the last decade of the nineteenth century and right through the twentieth century was through take-overs and mergers. Gradually, they began to dominate key industries – energy, tobacco, chemicals, steel, agribusiness, media, pharmaceuticals, aviation, PR, media and beverages.

The year 1886 marks an important milestone in the evolution of modern corporations. In that year the US Supreme Court, in the case of *Santa Clara County v. The Pacific Railroad Company*, ruled that the Fourteenth Amendment of the Constitution, which was designed to protect the rights of human beings, also applied to corporations. This gave corporations the natural rights of citizens.

But rights assume that there are corresponding responsibilities. The corporations never assumed their proper share of responsibility – given their economic strength and global reach – towards the well-being of human communities or the natural world. Working conditions were appalling in mines,

---

2. Ha-Joon Chang, *Kicking Away the Ladder*, 2002, Anthem Press, PO Box 9779, London, SW19 7QA, p. 86.

construction sites and factories. Corporations were seldom held responsible for plundering the natural world by cutting forests and poisoning rivers and the air. Corporations believed that the State had no role in monitoring their behaviour or holding them to account as long as they were not breaking the criminal law. They still believe that the State ought to guarantee their freedom to go about their business of making profits without any hindrance. They continue to exploit every legal avenue in order to avoid basic responsibilities towards humankind or the natural world. When the giant energy corporation Enron collapsed it turned out that they had not paid taxes for four or five years.

THE GROWTH AND POWER OF TRANSNATIONAL CORPORATIONS (TNCs)

After World War II, the balance of economic and even political power in the world shifted from territorially bound, nation-state governments to companies that roam the world with a single goal in mind – to maximise profits. Of the top hundred economic entities, over half are now corporations, not countries. For example, General Motors, the world's largest corporation, has an income similar to that of Austria which is twenty-first in the world's ranking for Gross Domestic Product (GDP).

The top 500 corporations control almost one third of global Gross National Product (GNP) and 76 percent of world trade. Their influence continues to grow. According to the UN Conference on Trade and Development (UNCTAD), TNCs controlled 17 percent of global GNP in the 1960s. This had jumped to 24 percent by 1984 and pushed ahead to 33 percent by 1995. Presumably this share has increased further since then. Another way to look at this growing control of the global economy by TNCs is to realise that the sales of the world's ten largest TNCs exceed the combined GNP of the world's 100 'least developed' (economically) countries. This includes all of Africa.³

Many states offer favourable tax holidays and special corporate tax rates to TNCs. Ireland has one of the lowest rates of corporate tax. In 2003 it is 12.5 percent. The Justice Commission of the Conference of Religious of Ireland (CORI) was one of the few organisations to recommend an increase in corporation tax in the run-up to the 2002 general election.⁴

The justification for lowering corporate tax is that the TNCs provide jobs. But even this notion is somewhat illusory. The number of people employed by the top 500 corporations is 18.8 million. This figure represents less than one-third of one percent of the world's population and, as we will see later, TNCs shed jobs even when they are very profitable. One way to stop corporations playing one country off against another in terms of environmental legislation and workers' rights would be to have a single corporate tax rate globally.

2002 AN *ANNUS HORRIBILIS*

2002 was an *annus horribilis* for the corporate world. Scandals at the giant energy company Enron, involving individuals who were both close friends and financial backers of President George Bush, surfaced late in 2001. Enron's demise was the biggest corporate collapse in history. Arthur Andersen, the firm that audited Enron's books, was convicted of obstruction of justice on 15 June 2002.

In June, WorldCom, one of the largest telecom communications companies in the world, was accused of the biggest accountancy fraud in history. Mr Scott Sullivan, the chief financial officer, boosted the company's profits artificially by $3.8 billion. He did this by counting some of the company's costs as capital investment.

3. *Financial Market Lobbying; A New Political Space for Activists*, 2002, The Corner House, PO Box 31137, Station Road, Sturminster Newton, Dorset DT1O1Y, UK, p. 2.
4. *An Agenda for Fairness*, CORI Justice Commission, Tabor House, Miltown Park, Dublin 6, p. 60.

Ms Anne Mulcahy, the chairperson and chief executive of Xerox, who assured the public, in May 2002, that 'the company is completely transparent and the investors know everything', was embarrassed one month later when it was revealed that Xerox had overstated its revenue by $1.9 million over the previous five years. Also in May 2002, the Securities and Exchange Commission (SEC) charged the accountancy firm Ernst and Young with entering into a joint venture with PeopleSoft while auditing the firm's books. In June 2002, Omnicom, the world's third largest advertising group, rejected allegations that it too had employed questionable accounting procedures. Its shares plummeted when its chief executive resigned in the summer of 2002.

Politicians, and regulatory agencies like the SEC in the US, condemned the greed of chief executives and promised tough action. In June 2002, President George W. Bush stated that he was 'surprised and outraged' by these revelations on numerous occasions. As Frank Rich explained in the *New York Times*: 'the corporate world has no reason to quake in its boots and submit itself to tough, independent regulations at national and global level. The reason is simple – corporations are now more powerful than many governments'. Rich's article, entitled 'All the President's Enrons', continued: 'Mr Bush keeps saying all the right things. He is "deeply concerned". He will "hold people responsible". But words, like stocks, lose value when nothing backs them up. It is now six months since the President promised "a lot of government enquiry into Enron". Since then *Playboy* has done a better job at exposing the women of Enron than the Bush administration has done at exposing the men.'[5]

In an effort to steady the financial markets President Bush visited Wall Street on June 9, 2002 to deliver a major speech on corporate accountability. True to form, the President

---

5. Frank Rich, 'All the Presidents Enrons', *The New York Times*, June 6, 2002.

talked tough but promised very little substantive reform. The editorial in the *New York Times* on June 10, 2002, entitled 'The Corporate Scandals', came straight to the point: 'at its core, however, the president's address was disappointingly devoid of tough proposals to remedy underlying problems in accounting, corporate governance and the safety net of federal laws and regulations that is supposed to prevent abuse.'[6]

The second editorial in the *New York Times* on that day raised serious questions about the President's own behaviour as a business man before being elected President. It stated: 'President Bush needs to speak much more frankly about the money he made in selling his faltering oil company to Harken Energy of Texas and later selling Harken shares shortly before the company's stock price collapsed.' The company headed by his Vice-President, Mr Cheney, Halliburton, is under investigation by the Securities and Exchange Commission; yet it secured lucrative contracts to rebuild Iraq after the 2003 war.

The political reality in 2003, on both national and global levels, is that corporations have little to fear from governments for the simple reason that they have colonised governments to such an extent that their vested interests take precedence over the rights of ordinary citizens.

We have no political institution commensurate with the global economic reality. Given the dominance of TNCs, the State is seen as an enforcer of the rules of the market, which will always favour the rich and powerful. In a truly just and sustainable world the economic system or systems ought to be subordinate to the political system. Without this, economic considerations become the yardstick to evaluate all other human activities. If something does not have value in the market-place it is seen as having no value at all. What is good for humanity is today defined in monetary terms, unlike the definition of human good in almost every other period of history or culture.

6. Editorial, *New York Times*, June 10, 2002.

TNCs AND POLITICIANS

The political power of TNCs was clearly evident in the aftermath of the US presidential election in 2000. It was obvious that the powerful corporations which had backed President Bush were looking for their rewards. The tobacco giant Philip Morris gave $2.8 million to the new President to fund his inauguration. In all, tobacco companies gave $7 million to the Bush campaign. In response to this it now seems certain that the federal government will abandon lawsuits against tobacco companies. This will save tobacco corporations about $100 billion in compensation. Looked at from a business perspective, this is a huge return on their investment.

Within a few weeks of his election, President Bush rescinded rules that would make mining corporations pay for the clean-up if they contaminated the public water supply. He also dropped safety limits on the levels of arsenic in public water schemes that had been imposed by the Clinton administration. Mining companies contributed $2.7 million to Mr Bush's campaign. At the same time, President Bush overturned a regulation that was aimed at protecting 60 million acres of national forest from logging. Logging firms contributed $3.2 million to the Bush campaign.

In November 2002, the Bush administration announced the most sweeping move for a decade to loosen industrial air pollution rules and further damage the US Environmental Protection Agency (EPA). The former head of the EPA, Carol Browner, described the changes as 'nothing but a special deal for the special interests and [it] comes at the expense of all who breathe and most particularly our children.' [7]

The oil industry is very close to President Bush. The Vice-president, Dick Cheney, worked all his life for the oil industry. In repudiating the Kyoto Protocol, President Bush was pandering to big oil companies. These contributed $25 million to

---

7. Mathew L. Wald, 'E.P.A. Says It Will Change Rules Governing Industrial Pollution. *The New York Times*, November 23, 2002, available, webpage http:www.nytimes.com/2002/11/23.

the Republican election war-chest and only $7 million to the rival Democrats. But it didn't stop there. In 2001 ExxonMobil wrote to the White House requesting that Robert Watson, the chair of the UN's Intergovernmental Panel on Climate Change (IPPC), be removed from office because of his strong stand on climate change. President Bush delivered for his friends even though the report he had commissioned from the National Academy of Science stated bluntly that global warming was real and getting worse.[8] Watson was fired in April 2002.

The biotech industry, which I will focus on in this book because of its plan to control the staple foods of the world, is also well represented in the Bush cabinet. The Secretaries of Defence, Health and Agriculture, the Attorney-General and the Chairman of the House Agriculture Committee have had connections with the biotech company Monsanto and other, similar corporations. The Attorney-General, John Ashcroft, received $10,000 (£6,800) from Monsanto in the 2000 elections – the biggest contribution that the company gave to any congressional candidate. He strongly advocated the promotion of genetically engineered (GE) crops in poorer countries during the past few years. The Secretary of Agriculture, Ann Veneman, was formerly a director of the biotech company Calgene. This company, which developed the first genetically engineered tomato, is now owned by Monsanto. Veneman has been promoting GE crops in world trade talks. Donald Rumsfeld, the Defence Secretary, was president of Searle Pharmaceuticals when it was bought by Monsanto.

Many other key players in the Bush White House have come straight from corporations. Mitchell Daniels, head of the White House Office of Management and Budget, is a former vice-president of Eli Lilly, a transnational pharmaceutical corporation.[9] Thomas White, vice-chairman of Enron

8. Robert Tait, 'Bush Warned to Act now to Curb Global Warming', *The Scotsman*, June 8, 2001, p. 23.
9. Julian Borger, 'All the President's Businessmen', *The Guardian* (Supplement), April 27, 2001 pp. 2 and 3.

Energy when it allegedly hid $500 million in losses and manipulated the California energy crisis, is Secretary of the Army.

The giant construction company Betchtel has very close connections with the Republican Party in the US. George Shultz, a former Secretary of State, was a board member of the company, as was Caspar Weinberger, who was President Reagan's Secretary of Defence. Betchtel which contributed generously to Republican candidates in the 2002 mid-term elections, has now won a contract worth around $680 million to rebuild schools, roads, hospitals and other installations which were damaged during the massive bombing campaign by the US and British forces in the 2003 war against Iraq.[10] The reason given for going to war was that Iraq had weapons of mass destruction and had the capacity to use these against their neighbours and even Western Europe. As of mid-September 2003, neither of the occupying forces – the US nor the UK – has been able to prove the existence of any credible weapons programme. The invasion did topple the murderous regime of Saddam Hussein, but the cost to the physical, social and cultural fabric of the country has been horrendous. Many feel that the real reason for the second Gulf War was to protect oil supplies for western corporations and customers and to give the US military more secure bases in the Middle East.

It is clear that war benefits a variety of armament industries. But there are other benefits. One example of this is what Paul Bremer, now the US-appointed governor of Iraq, did in the wake of the September 11, 2001, terrorist attacks in New York and Washington. One month later, he launched a company designed to capitalise on the fear of terrorism that had gripped corporate boardrooms in the US. Crisis Consulting Practice is a division of the insurance company Marsh and McLennan. Bremer's company specialised in integrated and comprehensive solutions to contemporary crises in the corporate world. These range from potential terror attacks to how

---

10. 'Bechtel to Rebuild Iraq', *The Ecologist*, June 2003, p. 7.

to deal with demonstrators who protest against the damage which globalisation policies are having on local inhabitants. In a paper entitled 'New Risks in International Business', Bremer admits that free-trade policies may undermine local economies and result in unemployment, and may have other negative social consequences; but he is adamant that it also leads 'to the creation of unprecedented wealth'.[11] His former company, Marsh, reaped its share of that new wealth in 2002. Operating income was up 31 per cent and the current chairman, Jeffrey W. Greenberg, foresees continuing gains in the coming years.[12]

TNCs ARE ALSO POWERFUL IN EUROPE

While the links between big business and politics are legendary in the US, people might feel that things are different in Europe. Many people who are favourably disposed to the enlargement and further integration of the European Union are unaware of the power that TNCs wield. The *Guardian* columnist George Monbiot revealed that a corporate lobby group called the European Round Table of Industrialists (ERT) has been the principal architect of European integration for the past seventeen years. ERT promoted the idea of the Single European Act quite a number of years before it was completed in 1992. The drafters of the Nice Treaty bowed to ERT's demands that national parliaments be bypassed in some trade negotiations. In future, the Commission will negotiate such international treaties.[13] The ERT has set its face against the new constitution being proposed for the EU and especially the proposal to have a President of the EU.[14]

Given this connection between politicians and corpora-

---

11. Naomi Klein, 'Bush's Can-do Man Puts the Business into Baghdad', *The Guardian*, June 5, 2003, p. 24.
12. Ibid.
13. George Monbiot, 'Stealing Europe', *The Guardian*, June 20, 2001.
14. 'Peter Sutherland's Homecoming', *The Phoenix*, June 6, 2003, p. 3.

tions, many people are worried about the concentration of the production and marketing of the world's food in the hands of a small number of biotechnology and agribusiness companies. George Monbiot bemoans the 'astonishing rapidity [whereby] a tiny handful of corporations are coming to govern the development, production, processing and marketing of our most fundamental commodity, food.' He believes that the power and strategic control they have amassed will make the oil industry 'look like a corner shop.'[15]

It is frightening to think that, within a few years, the world's food supply could be dominated by eleven or fewer giant TNCs. Agribusiness, biotech and pharmaceutical companies have increased their control through the provisions on intellectual property rights contained in the Uruguay Round of the General Agreement on Tariffs and Trade (GATT). We face the appalling possibility that, within a few years, the entire genome of many species, including humans, could be owned by a handful of companies and corporations.

These mammoth institutions are responsible for most of the world's resources extraction, as well as production processes, financial institutions, computer companies, and media. On the demand side, advertising and PR agencies like Burston Marsteller have created a consumer culture which is unsustainable. These institutions are not above using all sorts of methods to promote their wares. Everyone knows that cigarette smoking is a health hazard. Nevertheless, corporations like British and American Tobacco (BAT) used heavy-handed tactics to ensure that Third World countries opened up their markets to their products. If a country sought to ban BAT imports or impose steep health taxes, then, according to David Leigh, 'smugglers simply flooded the country with cartons of black-market cigarettes. The files reveal how BAT connived at these smuggling rackets, which were hidden in

---

15. George Monbiot, 'The Power of Transnational Corporations', *The Guardian*, September 17, 1997.

company reports behind such euphemisms as "GT" for "General Trade".'[16]

Corporations have huge budgets for lobbying politicias. In June 2003, confidential budget documents from some leading pharmaceutical companies showed that they plan to spend $150 million lobbying the US Congress and state legislatures, fighting price control around the world. They are prepared to subsidise 'like-minded' organisations and to pay economists for producing op-ed and other articles which promote the drug companies' agenda.[17]

TNCs also have intent on cornering the market in services – banking, education, water and sewage provisions, and power plants. In the Uruguay Round of GATT negotiations, giant US TNCs like American Express and Citicorp lobbied heavily for a separate section called the General Agreement on Trade and Services (GATS).[18] The service industry is big business, and dominated mainly by Northern TNCs. These corporations want to expand into the lucrative service-market but feel they are blocked at the moment since many services like health-care, education, transport, water, electricity etc., are provided by the state or heavily regulated by public authorities. Freeing up trade in these services will benefit corporations which are the driving force behind the new wave of privatisation schemes that is now spreading around the globe.

In the wake of the Doha meeting of the World Trade Organisation (WTO) in November 2001, the European Union has signalled that it will not be willing to dismantle its tradition of huge agricultural subsidies unless Third World countries open up their health, education, water, and transport services to transnational corporations. Immediately after Doha, the European Union set about targeting specific serv-

---

16. David Leigh. 'Clarke is not Fit to Lead the Tory Party', *The Guardian*, June 25, 2001, p. 14.
17. Robert Pear, 'Drug Companies Increase Spending to Lobby Congress and Government', www.nytimes.com/2003/06/01/national
18. George Monbiot, 'Gats Gaffes', *The Guardian*, March 9, 2001.

ices in countries as diverse as the Philippines, Paraguay, Peru, India, Malaysia and South Africa to be opened to foreign competition. By April 2002, the document setting these targets ran to 1,000 pages.[19] In India, for example, the EU wants the markets in waste management and water distribution to be opened up to foreign competition.

Once again, the poor will suffer. If TNCs are seeking to make profit out of public services like water, health and education then it is inevitable that those without money and purchasing power will lose out. Once governments have opened up the service section to the WTO there will be no going back. Democratic control over the decision on how to organise these services will be lost. GATS will cause particular havoc in the South.

LACK OF SOCIAL AND ENVIRONMENTAL ACCOUNTABILITY

TNCs have no loyalty to any people, country or place. Today they can locate anywhere in the world and operate efficiently from there. They are effectively accountable to no-one except their shareholders and are not held responsible for the social, cultural and ecological consequences of their actions.

My own town, Nenagh in County Tipperary, was on the receiving end of this kind of TNC behaviour in April 2002. Aventis Parma, the giant French pharmaceutical company, decided to close its factory there, with a loss of 230 jobs. Aventis had been in Nenagh for twenty years. According to the manager Alain Leduc, the move came as a result of a rationalisation programme which involved closing thirty-seven plants worldwide. The sickening thing for the workers was the fact that, according to Leduc, the Nenagh plant had been one of the top performing plants in the European region, operating to FDA and Japanese market standards. However, there were

19. Charlotte Denny and Larry Elliot, 'The Bananas for Banking Agenda', *The Guardian*, April 17, 2002.

other plants in Britain and France that would manufacture the products that were made at the Nenagh plant; so, savings could be made.

The decision was a body blow to the workers, many of whom have mortgages on their houses or are putting their children through third level education. The loss of 230 jobs is also a tragedy for a town of six thousand people. But none of this figured in the decision of the Aventis bosses to abandon Nenagh. The engine that drives it and every other TNC is the search for more and more profits so that the shareholders are happy and the stock market smiles benignly on the company. Everything else – the welfare of the community, the contentment of workers and the prosperity of the wider community, and, finally the health of the local ecosystem and the planet – are all secondary considerations of little consequence.

One might wonder whether Aventis was losing money and had its back to the wall, and so was forced to shed workers in order to remain profitable? Nothing was further from the truth. The turnover in 2001 for Aventis Parma, the world's largest drug maker, was €17.7 billion, and after-tax profits reached €1.6 billion, an increase of 40 percent over the previous year. But such a favourable balance sheet was not enough to save the jobs of people who had given twenty years loyal service to this company. They have simply been cut adrift to fend for themselves. [20]

With the power of national governments being rolled back in recent years, there are few avenues open today for checking the power of large transnational corporations. In fact TNCs want the national and global systems – economic, fiscal, social, cultural, environmental and political – to function for the purpose of providing a favourable climate for transnational investment and competition in the new global economy. John Vidal writes: 'so immense are they growing and such is their

---

20. Eibhir Mulqueen, 'Closure of Aventis Means Loss of 230 jobs', *The Irish Times,* April 13, 2002, p. 17.

skill in levering markets, so grand their resources and great their political influence that they are now effectively units of governance. Yet they have avoided, so far, the business of having to be socially and environmentally accountable, and are to all intents undemocratic and unaccountable.'[21]

The collapse of Enron towards the end of 2001 illustrates how the corporate world has structured itself to avoid any responsibility. The company soared from virtual obscurity to become a world player in the energy supply in a few years. It did this by adopting fraudulent accounting procedures which involved setting up around 900 shell companies around the world. In this way Enron was able to able to disguise heavy debts and losses. It also avoided paying federal taxes for four out of the previous five years.

International Herald Tribune columnist Richard Cohen captures the modern corporate ethos: 'Enron supported many charities and cultural institutions, but only the ones it chose. It put its name on a stadium, but, again only the one it chose. It basked in the gratitude it received for such largesse.'[22] It did not, however, pay taxes. Many of Enron's executives and board members made a fortune before the collapse of the company, while thousands of workers lost all their savings.

But what was done, while immoral, may not be illegal, since the laws are stacked in favour of the corporations. This point was made by two professors of accountancy – Michael H. Gandof and Stephen A. Zeff. During the 1970s and 1980s members of the US Congress, mainly from oil-producing states, brought pressure to bear on the Financial Accounting Standards Boards and the Securities and Exchange Commission not to demand tougher standards for financial reporting in the petroleum industry. The professors conclude that congressional involvement in financial standard-setting has been pure politics, fuelled by a system of campaign financing

---

21. John Vidal, 'Transnational Corporations and the Markets', *The Guardian*, April 30, 1997.
22. Richard Cohen, *International Herald Tribune*, January 22, 2002, p. 8.

that distorts the pursuit of the nation's legislative agenda.[23]

Congress facilitated corporate greed by designing the nation's laws in favour of corporations. What the chief executive of Enron, Kenneth Lay, affectionately named 'Kenny Boy' by President W. Bush, did may not have been illegal. It was, however, profoundly immoral. Enron was the single largest contributor to George Bush's political campaigns since he first ran for Governor in Texas. It also made a massive donation to Bush's presidential campaign. The Republican political analyst Kevin Philips traced the Bush family favour-swapping with Enron back to 1988 and likened Enron's potential damage to the Harding Administration's 'Teapot Dome' scandal. 'The question now is whether what went up together will come down together.'[24]

Very often the catch phrase 'get government out of business' is used as a smokescreen by the corporation to be allowed to do whatever is needed to make more profits. For example: at home in the US, Enron had campaigned vigorously to free energy companies from government regulations with the usual 'get the government off our backs' cry. But government help was eagerly sought for projects outside the US. In its drive to build power plants overseas and control the energy market, Enron benefited from a US$1.2 billion government-backed loan from two US agencies. In February 2002 the Overseas Private Investment Corporation (OPIC) was still owed US$435 million while the Export-Import Bank was due US$512 million. When it filed for bankruptcy in December 2001 Enron had approached OPIC for a further loan to bid for two major projects in Brazil. Enron did not tout the superiority of private enterprise when it was looking for what, in fact, is corporate welfare.[25] It is estimated that subsidies to

23. Michael H. Gandof and Stephen A. Zeff, 'Enron's Victims Can Blame Congress', *The International Herald Tribune*, January 24, 2002, p. 8.
24. Frank Rich 'All the President's Enrons', *The New York Times*, July 6, 2002, p. A13.
25. Associated Press Washington. 'Agencies Gave Enron US1$.2b in Loans', *Taiwan News*, February 26, 2002, p. 14.

corporations or corporate welfare, on a global scale, each year reaches $100 billion.[26]

TNCs DUBIOUS PARTNERS FOR DEVELOPMENT

During a preparatory conference for the UN Conference on Sustainable Development scheduled for Johannesburg in September 2002 many people from the non-governmental organisation (NGO) community were upset that First World countries seemed to have forgotten that crippling debt, inequitable trade and minimal development aid caused underdevelopment and needed to be addressed urgently. Many First World governments promoted the idea that TNCs ought to be seen as partners in delivering development goals. Many delegates from the South and members of the NGO community insisted that TNCs could not be seen as partners unless they were willing to become much more accountable for their activities on the human rights, social justice and ecological fronts.

TNCs were very evident at the Summit on Sustainable Development in Johannesburg in August 2002. Many were promoting themselves as environmentally friendly and as an essential element in overcoming global poverty and environmental degradation. TNCs would like people to forget their recent records on environmental and social issues. Key players in the present global economic system for over fifty years, many were involved in the fossil fuel business and TNCs have resolutely set their faces against any reduction in greenhouse gas emission. On the extractive side they logged tropical forests in unsustainable ways and promoted mining operations in unsuitable places. The agribusiness TNCs have made billions of dollars out of cornering agricultural activity globally and shaping it in a way that benefits the corporations rather than the farmers or consumers. Chemical companies have flooded the world with over 100,000 synthetic chemicals,

26. Thomas Berry, op.cit., p. 146.

many of which cause cancer or interfere with the endocrine and immune systems. Advertising companies have fanned the flames of consumerism in the First World and among the elite in the Third World.

Many NGOs were not impressed with TNCs' green rhetoric at the summit. Elizabeth Stuart of Christian Aid felt that the business community hijacked the meeting: 'What we are seeing is a history of business-friendly policies including self-regulation for corporate accountability.'[27] Friends of the Earth headed a coalition of environment and development groups at the Summit which were demanding that TNCs be regulated to prevent environmental and social abuse.

But the business community set its face against any external audit at both Rio and Johannesburg. Maria-Livanos Cattaui, the secretary-general of the International Chamber of Commerce rejected Christian Aid's proposals. 'Business would look askance at any suggestion involving external assessment of corporate responsibility, whether by special interest groups or UN Agency.'[28]

The Johannesburg Summit only briefly acknowledged that unbridled consumerism and a throwaway mentality is totally unsustainable and needs to be abandoned as quickly as possible if we are to avoid increasing poverty and environmental destruction. The Summit committed the international community to halving the existing number of people not connected to potable water supplies to 550 million by the year 2015, and to halving the number without proper sanitation to 1.2 billion by the same year. Such initiatives could save tens of thousands of lives. It is estimated that over 6,000 children die each day from diseases caused by poor sanitation and hygiene. But typical of the woolly nature of the Summit's thinking, there are no clear guidelines or effective strategies for achieving such laudable targets. In fact, according to *The New*

27. Barry James, 'Challenges of Development for Corporate Responsibility', *International Herald Tribune*, August 19, 2002, p. 5.
28. Ibid, p. 5.

Scientist individual countries and even corporations are 'left free to pursue approaches to managing water that are either wasteful or damaging to the environment.'[29] According to Jamie Pittock, water director of the World Wildlife Fund (WWF), 'summit agreements to improve water will not work if natural sources of water are not conserved and water used more efficiently.'[30]

Transnational Corporations would like to promote large building projects like dams and piping systems. The Summit played into the hands of the big building corporations, according to Torkil Jonch-Clausen of the Global Water Partnership: 'this summit has reduced the debate on water supply to arguing about money and pipes. There is no discussion about managing our river systems. It is a step back to the 1980s, before Rio ... It is a prime example of how the development lobby [transnational corporations] have snatched back the sustainable agenda from environmentalists.'[31]

As I mentioned earlier, TNCs, in the current climate of economic liberalisation, are targeting public services. A water consortium headed by the giant US construction corporation Bechtel Corp., took over the water system of Cochabama in Bolivia. Almost immediately it raised prices so high that people rioted and one protester was killed by the police. Feelings ran so high that the corporation's managers left the country and the service was returned to public ownership. Now Betchtel is suing the Bolivian government for $25 million.[32]

Why should we regard reducing the number of people without clean water by half to 550 million by 2015 as something to be proud of? Why not provide clean water for every person on the planet? The *Guardian* newspaper published a

29. *The New Scientist*, September 7, 2002, pp. 7-8.
30. Ibid p. 8.
31. *The New Scientist*, September 7, 2002, p. 11.
32. Barry James, op. cit. p. 5.

supplement in preparation for the Summit, entitled *Earth*. An article on 'Food and Trade' estimates that it would cost $170 billion to provide clean water and healthy sewage for all. Surely that should not be beyond the resources of our present global economy.[33] The Gulf War in 1991 cost $80 billion and we can be sure that the 2003 war against Iraq will cost at least twice that amount. The US occupation of Iraq is costing $1 billion a week. Why not divert the money wasted on arms to improving the quality of water and sanitation, health and nutrition for all the world's people?

GOVERNMENT SUBSERVIENT TO THE CORPORATIONS

As governments have succumbed to corporate pressure, economist Noreena Hertz claims that the ordinary citizen, either in the First World or Third World, has not been protected from the ravages of neo-liberal economics.[34] She accepts that political institutions in many Western democracies have been hijacked by the corporations and, by and large, do their bidding. She points out that, since the 1980s, business around the world has sidelined national governments.

Reviewing Hertz's book in *The Tablet* (August 4, 2001) Terry O'Sullivan writes: 'meanwhile those citizens, or at least those that could afford to, have in every sense bought into this new arrangement, happily exchanging their democratic birthright for a mess of consumerism.'[35]

Efforts to make TNCs more ecologically and socially accountable have been talked about since the 1960s. It was on the agenda of the UN Conference on Trade and Development (UNCTAD) in Santiago in 1972. That led to the UN Economic and Social Council (ECOSOC) setting up a UN Commission on Transnational Corporations as a research

33. 'Food and Trade, Earth, *The Guardian and Action Aid,* August 2002, pp. 38-39.
34. Noreena Hertz, *Big Business Buys Politics*, 2001, London, William Heinemann.
35. Terry O' Sullivan's review of Noreena Hertz's book, *Big Business Buys Politics, The Tablet*. August 4, 2001.

and administrative body. Almost immediately, a code of conduct for TNCs became a top priority for this body. In the 1980s, the US was in the grip of President Reagan's neo-liberal economic policies which believed that regulating TNCs was anathema. In March 1991 the administration of George Bush (Senior) requested all its embassies around the world to lobby against any further move on the UN Code of Conduct for TNCs. This lobbying was so successful that further attempts to get codes of conduct on other activities was effectively halted. A few had already been agreed – like the 1981 International Code of Marketing on Breast-Milk Substitutes that I will discuss later.

It is ironic that the demise of the UN Commission on TNCs coincided with the UN Conference on Environment and Development (UNCED) – dubbed the Earth Summit – in Rio de Janeiro in 1992. The pay-off for getting the corporate world on-side for the Summit was to emasculate the UN agency which had been attempting to draw up a code of practice for TNCs. The agency had its staff reduced, was transferred to Geneva and eventually disbanded.[36] The secretary-general of the Earth Summit, Maurice Strong, himself a well-know Canadian businessman who made his fortune in the energy business, invited the newly formed Business Council for Sustainable Development to write the recommendations on industry and sustainable development for Agenda 21 (UNCED's global plan of action). This was certainly a case of asking the foxes to guard the chickens.

TNCs abhor outside regulations or monitoring and have resisted them at every turn. But, in the light of recent scandals, corporations know that the heat is on and that it might be difficult to avoid calls for transparency and accountability. Even President Bush is talking a tough line, though whether he means it or not is open to question. Many TNCs have

---

36. David C. Korten, *When Corporations Rule the World*, 1995, Kumarian Press, Inc.,West Hartford, Connecticut, p. 375.

decided that the best way to avoid outside scrutiny is to promote self-regulation. Over the past few years they have been busy developing their own voluntary codes of conduct to show how sensitive they are to the human rights, social justice and environmental implications of their activities.

These voluntary codes, which cover just a few areas of commercial activity, have only been partially successful, and they are no substitute for external monitoring of corporate behaviour.

For example, organisations like the World Health Organisation (WHO) and the United Nations Children's Fund (UNICEF) and numerous NGOs have tried to stop the irresponsible marketing of infant formula foods by TNCs, especially in economically poor countries. The infant formula companies had used every marketing ploy to convince mothers to stop breast-feeding their babies even though all the medical evidence points to the health benefits of breast-milk. 'In the 1960s, health professionals working in developing countries pointed out the potentially fatal consequence of the inappropriate marketing of breast-milk substitutes – "commerciogenic malnutrition" as the director of the Caribbean Food and Nutrition Institute in Jamaica called it.' [37]

After protest campaigns by many NGOs, Church-based groups and the UN Agencies, a Code of Marketing of Breast Milk Substitutes was agreed in 1981. While undoubtedly it represented some progress and banned blatant propagandistic advertising many corporations, like Nestlé, continued to give free supplies of infant formula to health-care workers. In Russia, Nestlé gave the authorities their Russian-language translation of the Code which significantly reduced the impact of the agreed code. For example, it allowed the companies to advertise directly in maternity wards in hospitals.[38]

37. *Codes in Context, TNC Regulation in an era of Dialogue and Partnership*, February 2002, The Corner House, PO Box 3137, Station Road, Sturminster Newton, Dorset DT10 1YJ, UK, p. 2.
38. Ibid. p. 5.

Global and national regulations must be drawn up in such a way that producers and consumers carry their own full production costs rather than dumping them on other people and on the environment as has been happening during the past few decades. This was recognised by the 1999 Human Development Report from the United Nations Development Programme (UNDP) when it stated: 'Multinational corporations are already a dominant part of the global economy ... yet many of their actions go unrecorded and unaccounted ... They need to be brought within a frame of global governance, not just a patchwork of national laws, rules and regulations.'[39]

Commercial confidentiality is often used by corporations to defend themselves from public scrutiny. This must be drastically revised and should only be used to protect companies from industrial espionage. Otherwise access to their operations ought to be covered by normal freedom of information procedures.[40] In Chapter Five, I will discuss how three powerful agribusiness corporations used commercial confidentiality reasons for refusing to co-operate with the Mexican government when researchers discovered that genetically engineered maize had been accidentally planted in Mexico. Because Mexico is the home for hundreds of varieties of maize, genetic damage to that reservoir could have disastrous consequence for maize around the world, since it has now become a global crop. The arrogance of the companies involved – Monsanto, Syngenta and Aventis – must not be allowed to continue or go unchallenged, particularly when the stakes are so high.[41]

Mexico is not the only government that is afraid of taking

---

39. UNDP, *Human Development Report: Globalization with a Human Face*, Oxford University Press, Oxford, p. 100. Quoted in *Codes in Context: TNC Regulation in an Era of Dialogues and Partnerships*, February, 2002, p. 17.
40. George Monbiot, 'Turn the Screw', *The Guardian* (Supplement), April 24, 2001, p. 15.
41. Paul Brown, 'Mexico's Vital Gene Reservoir Polluted by Modified Maize', *The Guardian*, April 19, 2002, p. 19.

on giant pharmaceutical companies. During the past twenty years in Ireland alone, seventy-nine people have died and thousands of peoples' lives have been destroyed by HIV or Hepatitis C. They contracted these fatal and debilitating diseases through faulty blood products developed by transnational corporations. Many more will die from these conditions in the next few years.

It is still not clear whether the Irish government will be willing to pursue the offending corporations for compensation. The Minister for Health, Mr Martin, told the Dáil in November 2002 that 'the Government was likely to launch an inquiry into the actions of US-based multinational pharmaceutical companies implicated in the infection of haemophiliacs with HIV and Hepatitis C.' [42] Within two weeks, the Minister for Justice, Michael McDowell, was sounding a much more cautious note. He said that 'if people engaged in a multimillion lawsuit, then it must have a good chance of success. The Government should not simply rush into action to appear to be doing the popular thing when it could be a monumental waste of resources.' [43] The then chairman of the Irish Haemophilia Society, Mr Brian O'Mahony, criticised Mr McDowell and accused the government of lacking moral courage in its unwillingness to hold these companies accountable.[44]

Pharmaceutical TNCs act at times in irresponsible ways that prejudice the health of many people. In September 2001 thirteen of the world's leading medical journals, including *The Lancet*, *The New England Journal of Medicine* and the Journal of the American Medical Association mounted a concerted attack on pharmaceutical companies, accusing them of 'distorting the results of scientific research for the sake of prof-

---

42. Christine Newman, 'Haemophilia Society Criticises McDowell', *The Irish Times*, November 23, 2002, p. 2
43. ibid.
44. ibid

its.'[45] They claim that drug companies 'tie up academic researchers with legal contracts so that they are unable to report freely and fairly in the results of the drug trials.'[46] This extraordinary development, with such worrying implications for public health, should be investigated immediately by competent and well-resourced government agencies and the medical profession itself.

The chance of that happening in the present globalized world environment is close to zero. In today's world, TNCs are kings who are regularly wooed by governments and who dispense largesse to many doctors in the form of free trips to international drug company-sponsored conferences. This courageous intervention by the reputable medical press, though timely, received little media coverage. Pressure by corporations on researchers will further deepen the distrust that many feel about the reliability of in-company research trials, simply because billions of dollars are at stake for the company if a drug proves successful or has to be discarded.

Professional groups like doctors must set up transparent structures that distance them from pharmaceutical companies. In May 2003, the *British Medical Journal* (*BMJ*) called for 'relationships which are less grubby between companies and doctors.' The editor of the *BMJ*, Richard Smith, is critical of the manner in which drug companies fund and publish research. While recognising that drug companies need to advertise their products in the media and through sales representatives he feels that promotion offers which include 'free food' and 'a bag of goodies' are having undue influence on doctors and ultimately their patients.

Ray Moynihan, one of the contributors to the *BMJ*, believes that 'drug companies have got to learn how to win friends without buying them and doctors have to learn to value their

---

45. Sarah Boseley, 'Drug Firms Accused of Distorting Research', *The Guardian*, September 10, 2001, p. 2.
46. ibid.

profession's credibility without having to sell it.'[47]

Medical organisations need to put much clearer policies and protocols in place, based on sound ethical and professional criteria, to regulate the relationship between doctors, hospitals and drug companies. There must also be a much greater level of transparency in all these relationships and effective sanctions when individual doctors or drug companies transgress or subvert the policies. This means making resources available to monitor these interactions so the relationships are truly collaborative and not exploitative. One can only applaud and admire the stance taken by the *BMJ* and its editor. I am sure that the powerful British pharmaceutical companies are not happy about having their 'grubby' practices debated in public and that the the editor and the journal will come under huge pressure from them.[48]

One needs to be extremely cautious about the validity of research emanating from institutions or centres which receive large corporate donations. In 2001 Nottingham University received £3.8 million from British American Tobacco to establish a centre on corporate responsibility.[49] It hardly comes as a surprise that the researchers who published a paper in the *British Medical Journal* in May 2002 which found that passive smoking was not linked either to lung cancer or to coronary problems, received money from tobacco companies.[50]

THE NEED TO STRENGTHEN ANTI-TRUST LAWS

Global anti-trust laws should be enacted to ensure that corporations do not grow so large that they enjoy effective monopolies as Microsoft does in the computer world. It controls 95 percent of the software which is run on desktop

---

47. James Meikie, 'Medical Journal Turns on Drug Firms', *The Guardian*, May 30, 2003, p. 14.
48. ibid.
49. ibid.
50. ibid.

computers globally. During the anti-trust case taken by the US Department of Justice against Microsoft the government produced accusations that Microsoft bullied potential competitors out of the market. At the end of the trial in June 2000, Judge Thomas Jackson found Microsoft guilty of 'misusing its monopoly power and determined that it should be broken up'.[51]

It was fortuitous for Microsoft that the appeal should be heard during the new Bush administration. The US Court of Appeal, while it upheld a number of Judge Jackson's conclusions, criticised the judge's behaviour during the trial, struck down a number of his conclusions and, most important of all, did not direct that Microsoft be broken up. It is still possible that the US government will appeal this judgement but few expect this to happen as the new Bush administration looks very favourably on big corporate America.

Apologists for TNCs, among them the vast majority of governments, politicians and publications like the *Economist* magazine, claim that, in the long run, everyone benefits by capital markets being regulated, by privatisation and by other aspects of globalisation. A paper, 'The Emperor Has No Growth', challenges this accepted view and supports the claims of NGOs that globalisation has benefited the rich and hurt the poor. Researchers for this paper took the UNDP human development index as the basis of comparison rather than the more crude Gross Domestic Product (GDP) tables. They looked at two periods in recent history. These were 1960-1980, when governments intervened to protect their citizens; and 1980-2000, when many government proclaimed their faith in the market to solve social inequalities.

The researchers found that the period between 1980 and 2000 showed a very clear decline in progress. The poorest countries went from a *per capita* annual growth of 1.9 percent

---

51. Karlin Lillington, 'Government Change May Mean Lighter Penalty for Microsoft', *The Irish Times*, July 6, 2001, p. 11.

in 1960-1980 to a decline of 0.5 percent a year between 1980 and 2000. Things were even worse in the middle group of countries. They plummeted from 3.6 percent growth in 1960-1980 to just under one percent in the period 1980 and 2000.[52] Market forces alone will not produce economic growth or human and earth well-being. We need to curb the power of transnational corporations.[53] Strong anti-trust measures must be introduced in nation-states and in global institutions like the WTO.

There is an urgent need to renew the campaign for national and international agreements to monitor and regulate TNCs, so that they manage their economic activities in the service of the common good. At present, they have devastating impact on the air, water, soils and natural systems of the Earth. This cannot continue indefinitely in a finite world. There is a moral imperative to reform these institutions radically. Thomas Berry believes that the dominant profit motivation of corporations during the past 150 years will have to be replaced by a dominant concern for the integral life community.

---

52. Jonathan Steele, 'New Research Shows that Economic Growth Worldwide Has actually Slowed during the Era of Globalisation', *The Guardian*, August 3, 2001.
53. Thomas Berry, *The Great Work: Our Way into the Future*, Bell Tower, New York, 1999, p. 118.

TWO

From Agriculture to Agribusiness

IN 1956 an article appeared in the *Harvard Business Review*, which was destined to have a profound impact on how food is produced globally today. The author, John Davis, who later became Secretary for Agriculture in the Eisenhower Administration, wrote: 'the only way to solve the so-called farm problem, once and for all, and avoid cumbersome government programs, is to progress from agriculture to agribusiness.'[1]

Sixty years ago the average family farm was small and pursued a mixed form of agriculture. The farmer and, nearly always, his family were self-sufficient in food. Surplus production was traded, usually on the local markets. The market economy was somewhat peripheral to the farming since the community was largely self-sufficient and consumed a minimal amount of energy. Crusaders for private enterprise felt that this kind of operation was very inefficient. They proposed to bring farm-production and the marketing of agricultural products together under a single business umbrella. They argued that if this were done the wonders of science and of research technology could be harnessed in the interests of food production and processing, and that the consumer would benefit by having an abundance and variety of food available at cheap prices. Because everyone was expected to benefit from this transformation of agriculture, government and industry pursued the changes with vigour.

Sixty years later, this dream of full and plenty and cheap, nutritious food is turning into a nightmare. Petrochemical

1. Lawrence Geoffrey, *Capitalism and Countryside; The Rural Crisis in Australia*, 1987, Pluto Press, London and Sydney, p. 131.

agriculture is destroying land and water and polluting the air. The huge increase in the use of chemicals is causing extensive health and environmental damage. Mono-cropping is promoting the use of chemicals and undermining biodiversity. The organic farmer relies on crop rotation and combinations of various crops and insects, like ladybirds, to protect crops. The agribusiness farmer, on the other hand, is forced to rely on a battery of chemicals to protect the crop from insects and pathogens. This, of course, means huge benefits for agribusiness corporations; but it is not sustainable in the long run because it sterilises the earth and is based on fossil fuels which are finite.

Furthermore, small and medium size farmers, all around the world, have been pushed off their land; farm technology and research have ignored the interests of the small, organic farmer and concentrated instead on energy-inefficient technologies and machinery – to the benefit of agribusiness corporations and machinery manufacturers. Mixed agriculture has been replaced by specialisation in one or two crops; animal power has given way to petroleum-powered tractors, and the nutrients from animal waste have been replaced by petrochemicals.

The transition has been facilitated with credit from financial institutions. Paying the interest on this money and replacing machinery to meet each new phase in the farm technology revolution has ensured that farmers are sucked deeper and deeper into reliance on cash-crop agriculture, no matter what its impact is on the local social, cultural and physical environment. There has been a huge increase in both soil erosion and the pollution of the land, which ultimately affect human health and well-being. Small farmers have been forced to migrate to sprawling cities where they continue to experience hunger as they do not have enough money to buy sufficient food. In 1950 only 18 percent of the population of Third World countries lived in cities. By 2000 that figure had jumped to 40 percent.

We are living through a global clearance policy much more extensive than the one that cleared the Highlands of Scotland in the eighteenth century. When small farmers have been cleared, the large farmers and corporations who acquire the land no longer grow staple foods. Instead they switch to export crops like soybeans, sugar, coffee and even flowers. When the prime land in a country is devoted to export-oriented cash crops there is pressure on the surviving subsistence farmers in many Third World countries to farm in unsuitable areas.

Even in the United States the situation for farmers is far from rosy. The farming population has dropped from 6.8 million people in 1935 to 1.9 million in 1995. Farmers involved full-time in agriculture are fewer than the total prison population. Between 1993 and 1997 the number of mid-size farms in the US fell by 74,440. In the heart-land state of Nebraska many farmers are mortgaging their land in order to survive. Suicide is now the leading cause of death among US farmers: it is three times the national average.[2]

In France and Germany the farming population fell by 50 percent since 1978.[3] In Britain 20,000 farming jobs were lost in 2002.[4] In Ireland 670,000 people were involved in agriculture in 1926. Most of these farms were small to medium family farms. By 1991 the number of people involved in agriculture had dropped to 154,000. This represented only one percent of the work force.[5] The Census of Agriculture in 2000 showed a further 17 percent decline since 1991.

---

2. Katherine Ainger, 'The New Peasant Revolt'. *The New Internationalist*, January/February 2003, p. 10.
3. 'Open Letter from World Scientists to All Governments Concerning Genetically Modified Organisms', *Institute of Science in Society, www.i-sis,org.uk* p. 3, No 4, May 19, 2003.
4. Ibid.
5. Richard Douthwaite, *Short Circuit - Strengthening Local Economies for Security in an Unstable World*, 1996, Lilliput Press Ltd., Dublin, pp. 8-9.

THE TRUE COST OF CHEAP FOOD

Agribusiness corporations will claim that food was never so cheap or abundant. It seems to be cheap but only because much of the real cost of food is not reflected in the receipt we receive at the check-out counter in the supermarket. Agribusiness has managed to externalise many of the costs of food production to other sectors of the economy and the environment. Because they have become so dominant these companies can also manipulate the markets so that the farmer gets less and less for his/her produce. Fifty years ago the farmers in North America received between 45 and 60 percent of what consumers paid for food. Today it is a mere 3.5 percent. As Michael Pollan wrote in 1998: 'the real money in agriculture – 90 percent of the value added to the food we eat – is in selling inputs to farmers and then processing their crop'.[6] The same problems face European farmers. Katharine Ainger shows that agribusiness corporations 'traverse the planet, buying at the lowest price, putting every farmer in direct competition with every other farmer'.[7]

The true cost can only be appreciated when we factor in all the hidden subsidies, the billions spent on medical bills and the huge toll that industrial agriculture is taking on the environment. Many of these costs will be laid at the door of future generations. Very seldom is the transport and greenhouse gas cost of our food properly accounted for. The modern production process encourages long distance transport of food across countries and continents. Between 1965 and 1998 the international trade in food jumped threefold. Food eaten by most New Yorkers has travelled at least 1,500 miles. In Britain, the distance milk is transported has increased thirty times in the past forty years.[8]

6. Michael Pollan, 'Playing God in the Garden', *The New York Times*, *(Supplement)*, October 25, 1998. I am quoting from an article which was downloaded by the Irish Environmental Organisation VOICE, p. 5.
7. Katharine Ainger, art. cit., p. 11.
8. George Monbiot, 'Sins of the Superstores Visited on Us', *The Guardian*, March 1, 2001.

Impact on Human Health

In May 2003 British consumers began to hear horror stories about some of the food they consume. They learned that large food processors are bulking up chicken destined for public institutions like hospitals and schools with bits of beef, pig-waste and poultry skin. Hormones are regularly pumped into chickens. An editorial in *The Guardian* made the direct connection between intensive agriculture and poor quality food. 'The industrialisation of the food chain means that the search for ever bigger profits drives companies to seek cheaper ways to produce food'.[9] The editor was annoyed that the Food Safety Agency (FSA) in Britain is not pursuing and prosecuting the people involved in such scams. All the Agency is seeking is a labelling system that would inform the wholesaler, not the consumer, about how the food has been treated. The FSA is probably aware of the enormous power which the food processing companies wield today.

In the US, contamination of school-meat with salmonella is responsible for 1.4 million illnesses and 600 deaths each year. In response to this, in June 2000 the Clinton Administration ordered salmonella tests on ground beef used in the federal school lunch programme. The meat industry was totally opposed to any testing regime, claiming that testing was very expensive and the science behind it flawed. They were wrong: illnesses and death dropped by 50 percent after testing was introduced. In 2001, the Bush Administration, bowing to pressure from the meat industry, dropped the testing demand and allowed the industry to irradiate the food instead. It is not a coincidence that many of the large meat processors are Republican supporters.[10]

According to the US Environmental Protection Agency

---

9. 'Fowl Food: We Need to Know What We Are Eating', *The Guardian*, May 24, 2003, p. 23.
10. Conor O'Clery, 'Warning! Fast Food Will Damage your Health', *The Irish Times Magazine*, April 2, 2001, p. 24. A review of Eric Schlosser, *What the All-American Meal is Doing to the World*, 2000, Harmondsworth, Penguin Books.

(US EPA), organic arsenic compounds are extensively added to animal feed, particularly to poultry and pig production, in the United States. The reason given is that these compounds promote growth and control parasitic diseases.[11]

In addition, worrying levels of pesticide residue are found in fruit, vegetables and root crops. In Britain in 1998, potatoes eaten by the average consumer received twelve pesticide applications. Popular varieties are sprayed sixteen times during the growing season. Carrots are well dosed with an average of four insecticides, three herbicides and two fungicides. Some of these are organophosphates, highly toxic chemicals that affect the central nervous system.[12] Michael Pollan met a potato farmer, Danny Forsyth, in Idaho who felt obliged to use toxic sprays because fast food corporations want their potatoes to be blemish-free. If potatoes show any sign of *net necrosis* the batch will be rejected. Even though *net necrosis* is only a cosmetic defect he feels he has to spray his fields with toxic chemicals, including an organophosphate called Monitor. Forsyth admits that he will not enter a sprayed field for four or five days after spraying.[13]

A number of British families whose children were born without eyes are locked in a legal battle with the giant US transnational chemical company Du Pont. They believe that being exposed to the once popular fungicide, Benlate, while the women were pregnant caused the injury to their children. Concerns about the safety of benomyl, the chemical agent in Benlate, have been voiced by environmentalists for decades. In 1997 Du Pont had Benlate tested in an independent laboratory in Yorkshire. The researchers, carrying out tests on rats, found that a 'high' proportion of the chemical was drawn to the eyes. In recent months Du Pont has withdrawn Benlate from circulation because of mounting litigation costs.

11. Mike Ewall, *www.energyjustice.net/fibrowatch/toxic.html* May 26, 2003, p. 1.
12. Joanna Blythman,'Toxic Shock' *The Guardian Weekend*, October 20, 2001, p. 53.
13. Michael Pollan, op. cit., p. 5.

The litigation around the Benlate saga proves how powerful TNCs are in the US. In the 1970s the US EPA suggested to Du Pont that the fungicide ought to carry a warning label that it might cause birth defects, and that experiments with laboratory animals showed a significant drop in sperm count. However, Du Pont persuaded the EPA that the concerns were unfounded and, as a result, no warning appeared. This is a classic example of the corporate control of government agencies in the US. Furthermore, in the March 2002, court proceedings taken by parents of blind children against Du Pont, the judge who was hearing the case in West Virginia dismissed the scientific evidence of the independent testing agency as unsound just before the lawsuit began.[14]

The health impact of chemicals on the consumer is now becoming evident. Research published in May 2001 by doctors from Imperial College School of Medicine, under the leadership of Dr Simon Taylor-Robinson, reported a fifteen-fold increase in cancer of the bile duct In Britain between 1968 and 1996. Since the function of the liver is to detoxify the blood, the toxins that are isolated from food hit the bile duct. While the study does not point the finger at any source, the huge increase in chemicals in agriculture and food production over the past thirty years must be a primary suspect.[15]

The Centers for Disease Control and Prevention (CDC) in the US calculated that between 1970 and 1999 food-borne diseases in the US increased by a factor of ten. According to the US Food and Drug Administration (FDA), at least fifty-three pesticides that are considered to be cancer-causing, are present in many of our staple foods. Farmers are more at risk from handling toxic chemicals than ever before. A study by the US National Cancer Institute found that farmers who handle herbicides are six times more likely to develop non-

14. Rob Evans and David Hencke, 'Boost for Parents' Court Fight over Son Born without Eyes', *The Guardian*, March 25, 2002, p. 12.
15. Felicity Lawrence, 'We Get CJD and Bile Duct Cancer so Others Get Rich' *The Guardian*, May 28, 2001.

Hodgkin's lymphoma than the ordinary population.[16]

Additives are also a problem. In June 2002, when St Barnabas First and Middle School in Drakes Broughton, Worcestershire, in England banned food that contained any one of twenty-seven colourings and preservatives there was a dramatic change in the behaviour and concentration of many of the pupils. 'The teachers noticed that children who usually found it difficult to concentrate were much calmer and more able to get on with their work.'[17] The experience at St Barnabas confirms the findings of a study by the Food Commission, an independent watchdog, which found that food additives could lead to hyperactivity and tantrums in a quarter of children exposed to them. It analysed the effects of five different additives on 277 three-year-olds from the Isle of Wight. The children were given a drink containing the artificial colourings tartrazine (E102), sunset yellow (E110), carmoisine (E124) and the preservative sodium benzoate (E211).[18] Hyperactivity was much more evident among the children who ate food and drinks that had additives.

There is also an increasing concern about bacteriological contamination with salmonella enteritis which can cause diarrhoea and other complications. Intensive animal husbandry has increased the incidence of this disease. In 1994, Professor Richard Lacey, a microbiologist at Leeds University, in his book, *Hard to Swallow: a Brief History of Food*, estimated that in the UK one egg in five thousand is contaminated with salmonella.[19]

The over-use of drugs and growth hormones in animal feed is one of the factors that has contributed to the increase in resistance in many bacteria to different antibiotics. Antibiotics are often included in animal feed in order to increase the

16. 'Health Fatal Harvest, *The Ecologist*, November 2002, pp. 23-24.
17. Sarah Cassidy, 'Food Additives Ban Improves Pupils' Behaviour' *The Independent*, November 22, 2002, p. 12.
18. Ibid.
19. Quoted in Martin Khor 'Horrors and Dangers of Modern Animal Farms and the Reforms Needed', *Third World Resurgence,* No. 69, pp. 12-15.

weight of the animal. The situation is so serious at the moment that there is a real fear that current antibiotics are unable, in many cases, to deal with E.coli 0157:H7.

According to *National Geographic*, May 19, 2002 our fish, cattle and chickens are raised in giant factory-farms where animals are packed tightly into small areas. One picture in the magazine showed a single feedlot in Colorado with 100,000 cattle. Such large-scale agriculture keeps prices low but it also raises the risk of faecal contamination, disease and distress for the animals.[20] These practices need to be changed, and animals ought to be allowed to roam freely in the fields as happens in traditional agriculture in almost every part of the world.

Junk Food and Unhealthy Diet

Agribusiness is very much linked to the consumption of junk foods which are high in saturated fat, salt and sugar, and has a low nutrient content. This has led to an epidemic of obesity in many First World countries. Forty-four million US citizens are obese. Within that category there are 6 million 'super obese'. The rate of obesity in the US has doubled since the late 1970s.[21] By 2001, it was up 60 percent on 1991 figures. Obesity is also affecting children. In the USA, 13 percent of children aged between 6 and 11 years are overweight. In November 2002, it emerged that eight children in New York were suing McDonald's for failing to tell them that their daily diet of Big Macs and French fries would make them obese and, in some cases, make them diabetic.[22] A person who is 30 pounds overweight increases his or her risk of developing heart disease, diabetes, gall bladder disorders and arthritis. Overweight children are at risk of cardiovascular diseases, diabetes and other health problems.[23] Obesity is a factor in

20. 'How Safe?', *National Geographic*, May 19, 2002 pp. 9-31.
21. Conor O'Clery, art. cit., p. 24.
22. Oliver Burkeman, 'Youngsters Sue McDonald's for Failing to Warn that Fast Food Can Lead to Obesity', *The Guardian,* November 22, 2002, p. 19.
23. Conor O'Clery, art. cit., p. 25.

31,000 deaths each year and the cost to the British economy is over £2 billion.[24] World-wide, 46 million people die from cardiovascular and respiratory diseases, cancers, diabetes and obesity.[25] Obesity is also responsible for much back pain.

In Britain, one adult in five is obese, two-thirds of men and half of women being overweight. In Ireland, 39 percent of the population are over-weight and 18 percent obese, according to a study carried out in March 2001 for the Food Safety Promotion Board.[26] As one writer puts it, 'as soon as fast food outlets expand, waistlines soon follow'.[27]

Maeve O'Sullivan, a clinical nutritionist at Our Lady's Hospital for Sick Children, Crumlin, in Dublin, believes that childhood obesity is storing up problems for the future. Soft drinks are a particular problem in Ireland. According to her, Irish children are getting fatter because of poor diet, lack of exercise, and too many hours watching television. 'If you eat a fast food meal of a large burger, large fries and large soft drink, that would be over 2,000 calories, and you would need to run a marathon to burn it off.'[28]

There is no doubt but that fast food and soft drinks companies directly target children. In 1983 School District 11 in Colorado allowed Burger King to place advertisements in school hallways in return for educational funds. On an average day students will watch some thirty advertisements on television encouraging them to gorge themselves on fast foods or soft drinks.

One would think that governments and corporations would like to see measures taken to reduce this awful wastage of human life. Unfortunately, this doesn't happen because health

---

24. James Meikle, 'The Fat of the Land', *The Guardian*, April 24, 2002.
25. Gro Harlem Brundtland, 'Sweet and Sour', *The New Scientist,*, May 3, 2003, p. 23.
26. 'We're Getting Fatter', *The Irish Times* (Supplement). April 21, 2003, p. 25.
27. Anne Byrne, 'Ireland's McHappy Little People', *The Irish Times* (Supplement), April 21, 2001, p. 22.
28. Aideen Sheehan, 'Surge in Childhood Obesity a Timebomb for Health Services', *Irish Independent*, July 4, 2003, p. 3.

recommendations might lower corporate profits. In April 2003 the World Health Organisation (WHO) and the UN Food and Agriculture Organisation (FAO) published a document, *Diet, Nutrition and the Prevention of Chronic Diseases* which encouraged people to opt for a diet low in saturated fats, sugar and salt. Such a diet would include more vegetables and fruit, and people are also advised to take regular exercise.

The document took two years to produce and involved consulting sixty experts in nutrition and other related areas. Its recommendations did not come as a surprise. They simply spelled out the best current nutritional advice which is that people should limit their sugar intake. This did not please the powerful sugar industry. The US-based Sugar Association and World Sugar Research Organisation were so annoyed at the recommendation that sugar should provide no more than 10 percent of the energy in peoples' lives that they mounted a strong lobby in Washington. It attempted to discredit the report by claiming it was based on bad science.[29] Not content with this claim, the sugar barons want the US government to flex its political and financial muscles by stopping its $406 million per year contribution to the WHO.[30]

In 1999, Eric Schlosser, an investigative journalist with the *Atlantic Monthly*, published a book entitled *Fast Food Nation*. Not surprisingly, Schlosser found that fast foods are bad for one's health, and not just because of the saturated fats, sugar and salt content. He found that the taste one associates with French fries, sauteed onions or hamburgers at a typical fast-food outlet does not come from the potatoes, onions, or beef but from artificial, food flavouring. Much of it is supplied by factories in New Jersey such as International Flavors and Fragrance, the world's biggest flavour company.[31] We are eating food that is laced with chemicals, many of which may have adverse effects on our health.

---

29. Anne Byrne, art. cit.
30. Andy Coghlan, 'A Bitter-sweet Row', *New Scientist,* May 3, 2003, p. 13.
31. Connor O'Clery, art. cit.

Another reason for avoiding French fries as much as possible emerges from a study by the Swedish National Food Administration. It found that a cancer-causing chemical – acrylamide – is formed when starchy foods like potatoes are exposed to high temperatures. Because potato crisps and chips fall into this category, the Food Safety Authority in Ireland has warned consumers to cut down on such foods.[31]

In the light of such findings it is no surprise that in recent years a number of court cases have involved McDonald's, the largest fast food chain. While the company won the famous 'McLibel' trial, the revelations about practices in the restaurants during that two-and-a-half-year legal action humiliated McDonald's. Schlosser revealed that McDonald's was using beef tallow to make its French fries in different parts of the world. As a result, a member of the Jainist religion, Hitesh Shah, brought a case against McDonald's claiming that he and his fellow-believers had been misled: by eating French fries, they had transgressed a precept of their faith which prohibited them from eating beef.

Early in 2002, McDonald's settled by agreeing to pay $10 million to Hindu, vegetarian and other groups whose educational activities are closely linked to the concerns of Hindu and Jain consumers.[33]

Massive Soil Erosion

Another consequence of agribusiness is the extraordinary increase in soil erosion around the world. Poor land management, overgrazing, chemical agriculture, monocropping, deforestation and human population pressures have caused soil erosion and desertification on an unprecedented scale. The United Nations Environment Programme has estimated that since 1945 an estimated 108 million acres of productive land has been lost to agriculture each year. This adds up to

---

32. Aideen Sheehan, 'Consumers Urged to Lay off Chips after Shock New Cancer Discovered', *The Irish Independent,* April 25th 2002, p. 3.
33. Oliver Burkeman, 'Not so Big, Mac', *The Guardian, G2,* April 12, 2003, p. 2

4.85 billion acres, or around 35 per cent of the earth's fertile land.[34]

Professor David Pimentel and his team at Cornell University in Ithaca, New York estimate that, world-wide, about 75 billion tonnes of topsoil are lost each year at the cost of $400 billion. Most of this damage, unfortunately, is in the Third World where between 30 and 40 tonnes per hectare are eroded each year. Even in the US, 17 tonnes of topsoil per hectare are eroded with each cropping.[35] Topsoil is precious: without it, no crops will grow and pasture land will not be fertile. No machine can readily create topsoil. It builds up slowly; and it takes between 200 and 1,000 years for 2.5 cm of topsoil to build up. Erosion impoverishes the topsoil that remains: it loses nutrients and its capacity to retain water.

About 3500 million hectares, an area the size of North and South America, are affected by desertification. Each year, at least another six million hectares are irretrievably lost to desertification, and a further 21 million hectares are so degraded that crop production is severely affected. On April 18, 2001, scientists at the National Oceanic and Atmospheric Administration (NOAA) in Boulder, Colorado, reported a huge dust storm in northern China. Dust from that storm reached areas as far as Canada and Arizona, and obscured the foothills of the Rockies. During the past decade, similar storms have wreaked havoc on crop land and range land in China's north-western area. China's neighbours, Japan and Korea, are worried about the effect of these dust storms on their countries and are interested in working with the Chinese to combat the problem. The reason for the storms is that human pressure on the land in north-western China is excessive: the land cannot support the human and livestock population in the area. Officials estimate that 900 square miles of

---

34. Paul Hawken, *The Ecology of Commerce,* 1993, HarperCollins, New York, pp. 22-3.
35. Tim Radford, 'Wearing the World away', *The Guardian* (*Supplement*), March 9,1995, p. 8-9.

land are being turned into desert each year. This is in an area where the grass once grew as high as a horse's belly.[36]

CONTROL OF AGRICULTURE AND FOOD RETAILING CONCENTRATED IN FEWER AND FEWER HANDS

In the US, the control of food growing, processing, distribution and retailing has been concentrated in the hands of a few companies who now exercise a monopoly. According the Geoffrey Lawrence in his book, *Capitalism and the Countryside*, by the mid-1970s twenty corporations controlled poultry production in the US. Two oil companies increased the size of their beef-lots from 1,000- to 10,000-acre lots, in the Midwest, to 100,000-acre lots, in Texas. Three corporations – United Brands, Purex and Dud Antle (a subsidiary of Dow Chemical) – dominated lettuce production in California. Twenty-five of the giant supermarket chains accounted for over half of all US retail food sales. In California, 43 corporations owned 3.7 million hectares, more than half the farmland in the state.[37]

The situation in Britain is similar. Five chain-stores control 80 percent of grocery sales in Britain. This control, which amounts to a cartel, has been pushing down prices to farmers in recent years. Felicity Lawrence gives a good example of how this works.[38] She describes the plight of a farmer in East Anglia who grows peas. In 2002, he sold his produce to a processor for 17 pence per kilo. This is 8 pence less than he received in 1997. The processor charges 18 pence per kilo to clean, freeze and pack the peas. If the farmer or processor query these prices, they are told that the retailer can buy the peas much cheaper from an overseas supplier. The quality, of course, may not be the same. The supermarket sells these peas to the consumer for 98 pence per kilo. This is an enormous profit on

---

36. Lester Brown, 'Fear on the Wind' *The Guardian (Supplement)* May 20, 2001, pp. 8-9.
37. Geoffrey Lawrence, *Capitalism and the Countryside*,, 1987, Pluto Press, Sydney and London, p. 138.
38. Felicity Lawrence, 'Third Way to Poison a Food Chain' *The Guardian*, March 29, 200, p. 18.

a single kilo of peas.

The retailers are also attempting to push small and medium suppliers out of business in the name of rationalisation. Early in 2002, a leading supermarket in Britain called together its poultry suppliers and flatly stated that they wished to reduce the number to three. The lucky farmers had to reduce their costs and allow the supermarket to pay 'costs plus'. This means the cost of production plus a margin of profit agreed by the supermarket. So much for the working of the free-market!

SOCIAL IMPACT OF AGRIBUSINESS

Agribusiness has destroyed rural communities. During the 1980s, almost 100,000 farmers abandoned agriculture each year. The knock-on impact on the rural economies was huge. When agribusiness moves into rural communities the profits are not re-invested in the rural community but are siphoned off to the urban world or exported to the 'mother' country. A study by Walter Goldschmidt in California's Central Valley found that in the area where family farms predominated, the people in the surrounding rural towns had a higher standard of living, better community facilities, public services and amenities like parks and stores than in towns where corporate agriculture held sway.[39]

In Third World countries like the Philippines, there has been a huge shift from food production to growing cash crops. Almost half the agricultural land is devoted to growing luxury crops like pineapples and bananas for First World consumers. When profit is the driving force in agriculture, corporations will grow flowers or tropical fruit simply because they will make much more money growing carnations than growing food crops like rice or corn. This massive shift in land use is compromising the food security of many Third World countries whose people are, in turn, advised by organisations

---

39. Geoffrey Lawrence, op.cit., pp. 53-4.

like the World Trade Organisation (WTO) to buy cheaper cereals and other agricultural commodities on the global market.

The US, EU and Japan subsidise their farmers to the tune of $350 billion each year. Cotton production is a good example of how the livelihood of Third World farmers is being undermined by subsidies to First World agribusiness. There are 25,000 cotton farmers in the US, who receive an annual subsidy of $4 billion from the US treasury. This enables them to sell their product on the global market at a cheaper price than anyone else. In 2002, US producers controlled 40 percent of the global cotton market. This subsidy is wiping out 10 million cotton farmers in West Africa.[40] In the run-up to the G8 meeting in 2003 at Evian, the French President, Jacques Chirac, called for a moratorium on dumping cheap cotton, sugar and milk on African markets, as such dumping undermines the efforts of Third World governments to develop agricultural policies that promote food security. Food security demands that the staple crops of any country be grown locally.

ENVIRONMENTAL IMPACT

The world woke up to the environmental impact of modern agriculture when Rachel Carson published her book, *Silent Spring*,[41] documenting the impact of petrochemicals on land and rivers, and on bird life and insect life. Pressures to increase productivity and output have ensured that farmers are forced to employ techniques that are incompatible with environmental safety. Petrochemical agriculture does improve output over a short period, but at huge expense to the environment.

The use of pesticides is a short-term solution to the problem of crop pests. Insects' enormous gene-pool and their

---

40. Editorial 'G8 Summit Should Back Chirac's Plan', *The Guardian*, May 26, 2003, p. 12.
41. Rachel Carson, *Silent Spring*, 1962, Fawcett Crest, New York.

short reproductive cycles enable them to become resistant to even powerful chemicals in a short period of time. Pesticides also kill indiscriminately, even those insects that are beneficial to the farmer.[42] They also damage human health.

The dairy sector has also been radically changed in recent years. In 1970, the best dairy cows produced about 3,000 litres of milk per year. Today, thanks to breeding technologies and growth hormones, monster milkers produce 8,000 litres per year. In the United States a genetically engineered Bovine Growth Hormone (rBGH) is injected into dairy cows every fourteen days for 200 days of a cow's 335-day lactation cycle. This increases milk production but it takes a huge toll on the animals. Whereas in the 1970s many cows lived for ten years, the new super-milkers survive for only five.

Farmers also have to rely heavily on drugs to keep cows, with weakened immune systems, free from infection. According to Pascal Oltenacau of Cornell University, American and European dairy cows now need 100 drug treatments a year compared with 49 three decades ago to stay healthy.[43]

WILL INDUSTRIAL AGRICULTURE FEED THE WORLD?

Proponents of corporate agriculture claim that, without agribusiness, many more people would be hungry. I will address this 'feed-the-world' argument a number of times in this book, especially during my discussion of genetically engineered food. Here I will look at the claim that big farms are more efficient than small farms.

As Steven Gorelick points out in correspondence with Sean Richard in the *Ecologist*, the greater efficiency for larger farms refers to production per unit of labour.[44] Since small farms are more labour-intensive than larger mechanized farms, they

---

42. Paul Hawken, Amory Lovins, L. Hunter Lovins, *Natural Capitalism*, 1999, Little Brown and Company, New York, p. 196.
43. Richard Woods, John Elliot, and Jonathan Leake, 'The Terrible Cost of Our Cheap Meat'. *The Sunday Times* (March 18, 2001) p. 14.
44. Sean Richard and Steven Gorelick, 'Can Small Farms Feed the World?', *The Ecologist*, February 2001, pp. 20-3.

are less productive in that narrow sense. Studies by Peter Rosset of 'Food First', have shown that smaller farms are between 200 and 1,000 percent more productive than larger farms if one focuses on comparable units of land. Gorelick also points out the hidden social and ecological costs of agribusiness.

Similar conclusions were arrived at by researchers at the University of Essex in Britain. In 2002 they concluded the largest survey ever undertaken of modern sustainable agriculture in poor countries. The researchers found that, on average, sustainable farming methods yielded a 93 percent increase in per-hectare food production. For example: in Cuba in 1994, urban organic gardens only produced 4,000 tonnes of food. This grew to over 700,000 in a mere five-year period because of the rise both in the number of gardens and in their productivity.[45] The Prince of Wales believes that, if all the research money which at present is being ploughed into genetic engineering technology were channelled instead into improving agricultural methods that have stood the test of time, there would be a significant increase in crop yields without the problems associated with GE crops.[46]

GOVERNMENT POLICIES MUST SUPPORT ORGANIC FOOD

Claims and counter-claims about whether organic food is better for your health have been around for a long time. Petrochemical, intensive agriculture has always claimed that its products are as nutritious and as healthy as food produced in an organic way, without using petrochemicals. Regulators and government agencies have usually concurred. For example, John Krebs, the head of the British Food Standards Agency, has claimed that organic food is no better than food produced conventionally. However, John Paterson, a bio-

---

45. Jules Pretty, 'The Magic Bean', *The New Internationalist*, January/February 2003, p. 34.
46. The Prince of Wales, 'My 10 Fears for GM Food', *Daily Mail*, June 1, 1999, pp. 10 and 11.

chemist at Dumfries and Galloway Royal Infirmary, has criticised Kerbs for making such claims on the basis of very little information. He and his team at the University of Strathclyde have found that organic vegetable soups contain almost six times as much salicylic acid as food produced in a non-organic way. This acid is responsible for the anti-inflammatory action of aspirin. It also helps to prevent hardening of the arteries and bowel cancer.[47]

We ought to be aware of the variety of commercial and political forces ranged against organic food. These groups use studies from right-wing think-tanks like the American Hudson Institute and the Australia Institute of Public Affairs to discredit organic food and promote genetically engineered food.[48] The same point was made by Jim O'Connor in a letter to the *Irish Independent*. He claimed that most recent research from California shows that organic crops have 50 percent more cancer-fighting flavoniods than conventional food.[49]

EXTINCTION

Another consequence of agribusiness, with its promotion of monocultural cropping, is the huge loss of biodiversity in the world in recent years. When one thinks of the enormous amounts of food grown around the world annually, it comes as a shock to realise that global agriculture depends on a narrow species and genetic base. Of the 200,000 species of wild plants that are suitable for human consumption only a few thousand have been used and only a few hundred have been domesticated. More worrying still, three quarters of the world's food is derived from a mere seven species – potatoes, rice, wheat, maize, barley, cassava and sorghum. Nearly half of

47. "The Natural Choice: Organic Food Has More of What It Takes to Keep You Healthy', *The New Scientist*, March 16, 2002, p. 10.
48. Revd Paul Cawthorne, 'Organic Smear Campaign', *Green Christian Spring* 2000, Swiftprint, 3-5 Wood Street, Huddersfield, HD1 1BT. England, p. 11.
49. John O'Connor, 'Giving the Lie to Organic Detractors', *The Irish Independent,* April 22, 2003, Farming Section p. 5. See also his website www.plantorganic.com

the world's protein intake as food comes from rice, wheat and maize.

The genetic diversity within each of these crucial crops which, for example, will produce tall stalks in one situation and short ones in another, has been eroded as native habitats have been destroyed to enable industrial agriculture to opt for mono-cultured, high-yield crops. As a result, varieties with genes that protected crops from pests and adverse weather conditions are lost. The loss of such Farmers' Varieties or landraces, as they are called, is a biological disaster. When a crop with a uniform genetic structure succumbs to a disease – which happened to corn in the United States in the 1970s – researchers have to go back to the landraces to find varieties with genes that protect the crop against such a disease. The Canadian canola plant breeder, Percy Schmeiser, contrasts his plant-breeding with the genetically engineered variety of Monsanto. According to him there are two principal canola diseases in his area of Canada – stem rot and blackleg. He has bred a variety of canola which is resistant to these diseases. He can grow canola in a field for ten years without any problem. Monsanto, on the other hand, advise farmers to plant a field only every four years; otherwise, they will have problems with these diseases.[50] Whether it be corn, wheat, soya, canola or barley, farmers have developed varieties that are adapted to the soil and climatic conditions of the regions in which they live. The uniform varieties produced by TNCs like Monsanto need plenty of the company's own chemicals if they are to grow.

As more and more Farmers's Varieties are lost because of the aggressive corporate marketing for 'super' genetically uniform crops, our global staple crops become vulnerable to pests, disease and differing weather conditions. For example, India until a few decades ago cultivated over 30,000 varieties of rice. Now it is on the threshold of opting for one so-called

---

50. Interview with Percy Schmeiser, *WorldWatch*, January/February 2002, p. 10.

'super' variety. In the process, centuries of botanical knowledge and breeding techniques, which are our global food insurance policy for the future, are being lost. The corn failure in the US did not lead to hunger and starvation. In Ireland in the 1840s, the failure of the potato crop did lead to a disaster. Over a million people lost their lives and many fled to Britain, the US, and Australia to avoid the famine caused by the failure of a single genetic strain of a food crop.

Many of the international centres in which seeds are being stored and preserved, like the International Rice Research Center in Los Banos outside Manila, are suffering because of government budget cuts. However, the place to preserve biodiversity is not in university laboratories or research centres but by planting the crops in different environments. Seed banks can only survive for a decade or so. We need to get these traditional varieties back into the land.[51]

The Green Revolution provided agribusiness corporations with a golden opportunity to benefit from and control agriculture. This control is poised to increase dramatically with the introduction of genetically engineered seeds which are patented. Instead of promoting this dangerous technology, governments ought to be promoting more sustainable forms of agriculture.

The challenge of meeting food security could easily be addressed by organic agriculture. As the Prince of Wales said in his Reith lecture in May 2000: 'if a fraction of the money currently being invested in developing genetically manipulated crops were applied to understanding and improving traditional systems of agriculture, which has stood the all-important test of time, the results would be remarkable.' [52]

SCRIPTURE AND THEOLOGY

Finally, Christians ought also to be concerned about the

51. Paul Hawken et al., op. cit. p. 194.
52. The Prince of Wales, 'We Must Go with the *Grain* of Nature' *The Times,* May 18, 2000.

abuse of land which is at the heart of petrochemical agriculture. It is really soil-mining rather than stewardship. In our Judeo-Christian tradition land is one of God's most precious gifts to humankind. The second account of creation in the Book of Genesis tells us that God's involvement with humans does not end with creating them. Immediately, 'the Lord God planted a garden in Eden, in the east and there he put the man he had fashioned' (Genesis 2:8). God instructed the man to *till* and to 'keep' the land (Genesis 2:15). The Hebrew words used here have overtones of service – protecting and defending the land from harm. The tradition of stewardship has emerged from this perspective. Every seven years land was to be allowed to remain fallow in order to regain its fertility (Exodus 23:10-11). The cultivators were only God's tenants and it was clearly recognised that there were restrictions on what they could do with the land. 'The Land must not be sold in perpetuity, for the land belongs to me and to me you are only strangers and guests' (Leviticus 25:23).

It is clear from the Gospels that massively inequitable land distribution in Israel at the time of Jesus caused major suffering, poverty and dislocation. Powerful landlords, many of them associated with Herod Antipas, had taken over much of the land just as agribusiness is doing today. In his own life-time Jesus would have seen the livelihood of many independent land-owners being eroded. Some were reduced to tenancy and others took to the roads with bandit groups. Jesus' critique of the greed and acquisitiveness of the group knows as the Herodians was a direct response to their rapacious behaviour. He would also have known that the poor had to carry much of the costs of Antipas's building programme in Sepphoris (6 km from Nazareth) and Tiberias on the shores of Lake Galilee.

At the same time, the Temple cult was controlled by the Sadducees and Pharisees. These two groups, mainly centred in Jerusalem, grew rich through the tithes and various offerings which the poor gave when on pilgrimage to Jerusalem. This religious elite held something of a stranglehold over the way people were allowed to experience God at that time. Jesus' proclamation, grounded in his own experience of God, as 'Abba' who 'loved the world so much that he gave his only Son, so that everyone who believes in him may not be lost but may have eternal life ' (John 3:15-17) would have been experienced as liberation by the disciples. The elite saw it as subversion that had to be severely dealt with.

Jesus' own mission was directed towards these poor people in Galilee. He set out to inspire them with a vision of a new order grounded in his experience of God's care, goodness and love, and a just social order based on God's unconditional love. He reminded his listeners that the God who looked after the little creatures would also look after them (Matthew 6:25-35). Scripture scholar, Professor Seán Freyne, argues that Jesus was not a country philosopher adept at presenting universal truths to his audience. In fact it is his concern for the local situation of his time which makes his witness relevant for today.[53]

Finally, by linking his mission to the Jubilee (as he does in Luke 4:18 - 19) Jesus was making it very clear that he was on the side of the dispossessed:

> The spirit of the Lord has been given to me,
> for he has anointed me.
> He sent me to bring good news to the poor,
> to proclaim liberty to captives
> and to the blind new sight,
> to set the downtrodden free,
> to proclaim the Lord's year of favour.

Even though the Jubilee idea was never fully implemented in practice, at least it held out the hope that lands which had been appropriated, by a variety of means, would be returned

---

53. Seán Freyne, *Texts, Contexts and Cultures,* 2002, Dublin Veritas, pp. 201-2.

to the rightful owner after fifty years.

As in Jesus' time, economic pressures for the past fifty years, have pushed small farmers off their lands into endemic poverty in the sprawling megacities of Latin America and Asia. There, they lose their culture, social bonds and, very often, their dignity and self-worth. As a missionary, I know that hundreds of millions of displaced country people live in squalor in many large cities without access to clean water or sanitation. The lands which they have lost – stolen fields as they are often called in the Philippines – have been turned over to large landlords or agribusiness corporations. Unlike the subsistence farmers who cared for the land, the corporations see it as an economic asset to be exploited for maximum profit. When the soils are exhausted in one place they will move on to another location somewhere else around the globe.

Given that the Gospel was first preached in a very exploitative and unjust socio-economic context, it is amazing that the Churches today have very little to say about how, over the past fifty years, a small elite group has gradually taken over the lands of the world. Joseph Gremillion's massive volume *The Gospel of Peace and Justice*[54] which contains many crucial documents on Catholic social teaching from 1880 to 1975, has only a few passing comments about agriculture and land, apart from the encyclical *Mater et Magistra*.

FRUIT OF THE EARTH AND WORK OF HUMAN HANDS

Fruits of the land are at the heart of our Eucharist Celebration. Each time we celebrate the Eucharist we bring bread and wine to the altar, we recognise that it is 'fruit of the earth and work of human hands'. Bread comes to us from the fertility of the land, the skill and hard work of the farmer, miller and baker. This 'fruit of the earth and work of human hands'

---

54. Joseph Gremillion, *The Gospel of Peace and Justice*, 1976, Maryknoll NY, Orbis Books.

sequence is the ideal relationship that ought to exist between humans and the rest of creation. It is based on a mutually enhancing relationship rather than on one which exploits people and destroys land and human communities – as so often found happens in industrial agriculture.

In the Eucharist we bring gifts from our land, bread and wine, that are destined to be transformed into the Bread of Life and the Cup of Eternal Salvation – the Body and Blood of Christ, our ultimate source of nourishment as Christians. If we continue to pollute our lands these precious gifts will also be contaminated with the fingerprints of sickness and death. How, we might ask, can tainted wheat ever become the Bread of Life for us?

THREE

The ABC of Genetic Engineering

BEFORE EXAMINING the ethical issues involved I will outline, briefly and in a very simplified way, the various steps involved in genetic engineering.

Genetic engineering is a by-product of the relatively young science of genetics. The science emerged out of the pioneering work of the Austrian Augustinian priest, Gregory Mendel (1822-1884). In a paper published in 1865, *Experiments with Plant Hybrids*, Mendel developed a theory of organic inheritance from his work on crossing garden peas that exhibited different characteristics – like tallness or shortness, the presence or absence of colour in the blossoms (green or yellow), round or wrinkled appearance, and so on. In crossing the tall with the short variety of peas one might expect that the offspring might be of medium size. This is not so. Mendel discovered the dominance of certain 'traits' or dominant genes (as we now call them) over other genes that are called recessive. He postulated that in succeeding generations offspring appear to exhibit dominant and recessive genes in a specific ratio of 3:1. For example, sometimes the specific characteristic of shortness appears in one generation and is not found in the next generation but appears again in the third generation.

Mendel's experiments and laws help us to understand what is happening. According to his insight, when a recessive gene is paired with a dominant gene, the recessive gene cannot express itself and therefore will have no perceptible effect. In the next generation, however, the recessive gene of shortness might be paired with another recessive gene and therefore it will express itself and have a perceptible effect, producing a short plant.

Unfortunately for Mendel, neither his experiments nor the laws he formulated were appreciated in his own day. Mendel sent a copy of his 1866 lecture to many of the famous scientists of the day, including Charles Darwin. The article's pages remained uncut; so, it was clear that Darwin had not read Mendel's work.[1] If Darwin had read Mendel's article he might have understood better the process of evolution. It was not until the early 1900s that various other scientists, like Hugo De Vries, William Bateson and Karl Correns, independently obtained results similar to Mendel's. When these scientists checked the literature they found that Mendel had experimented with peas and published a theoretical understanding of the data thirty years previously.

By the 1920s, genetics was being used to help plant breeders improve the performance of agricultural plants. In the 1930s scientists began to explore the similarities between quantum physics and genetics. Erwin Schrodinger's book *What is Life?*, based on lectures he gave in Dublin, and published in 1934, brought fresh insights from modern physics and biology to bear on genetics and so gave impetus to new research. On the negative side, genetics got a bad name in the late 1930s and 1940s because scientists and doctors who were promoting eugenics in Germany, Nordic countries and the United States often appealed to genetics to bolster their arguments.

Even a short account of the history of genetics would be incomplete without mentioning the work of the great Russian geneticist Nikoloi Vavilov. He and his co-workers travelled all over the world between 1916 and 1940, studying crop diversity and establishing seed centres in the USSR. In a small area of Ethiopia he found hundreds of varieties of wheat which are unique to that place. Vavilov used his knowledge of genetics

---

1. Robin Marantz Heniz, *A Monk and Two Peas*, 2000, London, Orion Books Ltd, p. 140.

to pinpoint the places in the world where humans first cultivated particular wild crops like wheat, maize and potatoes. He also mapped out the areas in the world which are rich in biodiversity like the islands in South East Asia and Peru.

During the period from 1800 to 1950s scientists had learned a lot about living organisms. They knew that natural organisms, including the human body, were made up of cells that formed our blood, tissues and other organs. Robert Hooke, an English biologist, had already discovered and described the cell almost 350 years ago. Scientists knew that the cell was the simplest unit of life. It is estimated that the human body contains 60 thousand billion cells. Every cell is enclosed by an envelope called the plasma membrane. The cell contains a watery liquid called cytoplasm (this word is derived from the Greek word *kytos* meaning hollow and *plasma* meaning vessel) where the nucleus and smaller structures, called organelles, are found. The cytoplasm must be able to draw in sufficient nutrients from the environment and dispose of waste matter. The nucleus of the cell contains several chromosomes. A dog cell has 39 pairs of chromosomes while every human cell contains 23 pairs of chromosomes or 46 in all. Each chromosome is a long coil of DNA (deoxyribonucleic acid) rolled on itself and often represented in the form of a cross. Since the gene sequencing in the human genome project has been completed in 2002, we know that it contains around 35,000 genes (the totality of all genetic material of a cell or organism). Each gene is a tiny segment of DNA and carries information or is 'encoded' to instruct the cells to make the specific proteins which are the building blocks of life. These include blood, tissues, nerves, bones etc. Genes and proteins determine the physical characteristics of a living organism, like the colour of one's eyes or the height of a plant.

Much of the above information was known by 1950. In 1953 the science of genetics took another leap forward. Two young scientists at Cambridge University, James Watson and Francis

Crick, discovered the physical make up of deoxyribonucleic acid (DNA), the fundamental molecule of life, and produced a model using wires and cardboard. They discovered that the structure of DNA was like a double helix. The two strands were twisted around each other like a spiral staircase with bars extending across the connecting strands. These units, composed of four different chemical nucleotides, arrange themselves in a variety of patterns that form the genes. It is the precise ordering of the chemical base in the DNA molecule that makes each life form unique. In the light of Watson's and Crick's discovery, biologists began to realise that they could change or modify life forms.

But this discovery, though crucial, was not sufficient to enable scientists to cut up, delete, or recombine genes. They needed tools to cut the genes, and then they required a suitable mode of conveyance or vectors to insert the genetic material into another organism. The cutting tools were discovered in a group of enzymes that are called 'restriction enzymes'. They have the ability, as part of their own defence mechanism, to splice up DNA. The vector is also composed of genetic material like specific types of viruses or plasmids from bacteria. The function of the vector is to carry the genetic element across the cell membrane of an embryo of the host organism in order to introduce the foreign gene into the target cell's genome. This is exactly what recombinant DNA means.

The first genetically engineered organism appeared in 1973 when Drs Stanley Cohen and Annie Ghang inserted genes from a South African clawed toad into a bacterium – e-coli. When the e-coli reproduced it also reproduced the toad gene that had been inserted into the bacterium.

SOME TECHNICAL ELEMENTS OF BIOTECHNOLOGY

First, the geneticist has to identify the gene or gene sequence which expresses the desired trait. Next, a carrying agent or vector is needed whose job it is to carry the desired gene into the target genome. The vector is often a plasmid. This is a small DNA ring common to bacteria as additional DNA other than from their own chromosomes. The most commonly used is the Ti-plasmid for the bacteria *agrobacterium tumefaciens*. Several genetic elements are spliced together within the vector. At this point, we have the desired gene or gene sequence, the vector and the promoter which will enable the new gene to become expressed in the target organism.

Several techniques are used to transfer the new genetic material to the target plant. One way is for the Ti-plasmid from a pathogenic bacterium to infect the cells of the target plant. Another way is through a process called electrophoresis. This involves using an electric field to force the new genetic material into the target plant. A third way is to coat thousands of gold bullets with the genetic material, load them into a gene gun, and fire them into the cells of the target plants.

The final operation in the genetic engineering process is to determine whether the target plants have taken up the new genetic material. Until recently, the most common way of doing this was to use antibiotics to kill the cells that had not integrated the foreign gene. Cells that have taken up the foreign gene survive because they possess the resistant marker gene which will produce the protein for which it is coded. Finally, the biotechnologist completes the construct with a terminating or regulatory gene.

The biotech company Monsanto has genetically engineered soybeans to be resistant to their own Round-up Ready herbicide. The gene construct comes from the soil bacterium *agrobacterium sp*. The chloroplast targeting sequence is from the wall cress. Chloroplast is a plasmid containing chlorophyll and other pigments occurring in plants that carry out photosynthesis. The promoter is from the figwort mosaic virus and a terminal sequence is from the common pea.

When we look at the four steps described above, it is easy to understand why Dr Mae-Won Ho and others consider genetic engineering to be imprecise, inefficient and potentially dangerous.[2] She argues that genetic engineering is so crude and imprecise as to be inherently hazardous to health and biodiversity. The insertion of a foreign gene into the host genome is a complex and random process, not under the control of the genetic engineer. It is through the use of artificial vectors that the horizontal gene transfer is achieved. The transferred gene can give rise to random genetic effects, including cancer. Michael Pollan insists that a great many factors influence whether, or to what extent, a new gene will do what it is supposed to do. In one early German experiment, scientists succeeded in splicing the gene for redness into petunias. Everything was working out fine until the weather changed and became very hot. Suddenly and inexplicably, the petunias lost their red pigmentation.[3]

Five hundred and fifty-eight scientists, including the biologist David Bellamy and geneticist David Susuki, have written an open letter to governments telling them that, in their professional opinion, the technology is unstable. They argue that it will contribute to an increase in the frequency of horizontal transfer of those genes that are responsible for virulence and antibiotic resistance, and allow them to recombine to generate new pathogens.[4]

After the Philippine government had agreed, on December 16, 2002, to allow Monsanto to plant Bt corn, eighteen doctors and medical professionals wrote to the government.

2. See also Mae-Wan Ho, 'The Unholy Alliance', *The Ecologist*, Vol. 27, No. 4 (July/August 1997) p. 156.
3. Michael Pollan, 'Playing God in the Garden', *New York Times* (Supplement), October 25, 1998, p.3. I am quoting from a text that was downloaded by the Irish environmental organisation VOICE.
4. 'Open Letter from World Scientists to all Governments concerning Genetically Modified Seeds', *Institute of Science in Society*, www.i-sis.org.uk/ 17/05/03, p. 4.

They pointed out that the forced insertion of a foreign gene randomly into the genome of another species is itself inherently dangerous. The effects are unpredictable and largely unknown. Gene constructs could eventually insert themselves into the mammalian cells and even the human genome. It could disturb the gene ecology and could possibly disrupt important functions of the organism that could lead to abnormalities and/or diseases, including cancer. The effects are extremely difficult to predict. Existing risk assessment methods are largely inadequate and inappropriate.

Such fears are dismissed by other geneticists. But if there is even a remote chance of any of these side-effects happening the whole genetic engineering enterprise should be put on hold until independent scientific research has addressed the issues over a considerable period of time.

The mechanistic bias of conventional genetic engineering – the assumption that each gene articulates for a particular trait – took a bit of a hammering when the mapping of the human genome project was completed. Many scientists had predicted, on the basis on the number of proteins humans must synthesize in order to be human, that humans would possess 100,000 genes. Scientists were surprised when they discovered only 30,000 genes. It now appears that the mechanism of gene expression is much more complex and complicated than has been assumed in orthodox genetics. In fact we know that proteins are not made directly from the instructions on the DNA. What happens is that the relevant gene is copied initially on to a short-lived nucleic acid call RNA. This molecule then provides a kind of template on which the protein is built.[5] Yet genetic engineering still operates with the belief that a single trait can be transferred in a rather simplistic way.[6]

   5. Philip Ball, *H2O: a Biography of Water*, 1999, London, Orion Books Ltd, p. 208.
   6. Barbara Kingsolver, 'A Fist in the Eye of God', *Mother Earth News*, Ogden Publications Inc., 1503 SW, 42nd St, Topeka, KS 66609, pp. 51-8.

GENETICALLY ENGINEERED ORGANISMS
– SOME EXAMPLES

Scientists can re-engineer the genetic blueprint of an animal or plant in order to create a 'super animal' or a higher yielding plant.

A company called Calgene developed a genetically engineered tomato called Flavr Savr in the US and Europe. This tomato was approved by the US Food and Drug Administration (FDA) in May 1997. The goal of the experiment was to extend the shelf life of the tomato that is marketed under the brand name of 'McGregor'. Calgene invested a whopping $95 million in the process which involved isolating a gene that encodes for an enzyme involved in the ripening process. Having discovered the enzyme, the technologist blocked its expression. As a result, the tomato will take a few days to ripen on the vine and still maintain its firmness during shipping. As a consequence, the unfortunate consumer will be duped into believing that the tomato is much fresher than it actually is. Extending shelf-life means that the crop can be grown much further away from the retail outlet. For example, it is now possible to grow tomatoes in places like Central America where the labour and environmental laws are much less strict than those in the United States. Calgene had already been in touch with Mexican growers with a view to producing Flavr Savr on their lands.

Scientists have also attempted to create leaner and more cost-effective pork through genetically engineered pigs; one result was that the animals experienced extensive arthritis.

On the medical side, genetic engineering has created the first patented mammal, called the Onco Mouse. This creature was genetically engineered with a human gene to express cancer in its mammary glands so that it could be used for diagnostic and therapeutic procedures for breast cancer in humans. These are just a few examples of the wide range of transgenic plants and animals that are now available.

Today plants and animals with genes taken from completely unrelated species are being engineered in the laboratories of biotechnology companies and released into the environment. There is genuine worry about genetic pollution. In October 2001, researchers from the University of California found that one of the oldest varieties of maize in the world had been contaminated by genetically engineered maize in Mexico. This is a very worrying development as Mexico is the genetic home of maize. The UN Food and Agriculture Organisation is worried that genetically engineered crops may pollute the gene pool of conventional relatives in various parts of the world.

FOUR

The Pros and Cons of GE Food

BIOTECH CORPORATIONS use a number of arguments to promote genetic engineering. Firstly, they argue that genetically engineered crops will feed the world and banish hunger. They also claim that genetically engineered crops require fewer herbicides and pesticides than conventional crops and that using fewer petrochemicals will benefit human health and the environment. They go on to insist that genetically engineered crops are not really that new and, in fact, are in continuity with traditional breeding procedures. According to them, genetic engineering is only a more refined method for improving previous plant breeding technologies. Finally, they accuse those who are opposed to genetic engineering of being anti-science, modern-day Luddites.

I will begin by examining the claim that genetically engineered organisms are in continuity with traditional biotechnology. Genetic engineering entails the manipulation of an organism at cellular level to produce new, altered organisms with whatever traits are desired by the breeder. Genetic scientists cut out bits of a living organism's DNA genes, and splice them into a totally unrelated species. As we have seen above, animal genes are spliced into plants, bacteria genes are moved across to food crops, and even human genes are used to change animals and plants. The imported genes can destroy or influence the activity of other genes so that a completely new organism is created whose responses in a particular environment are unpredictable.

Traditional forms of biotechnology have been around for thousands of years. These include the processes of making beer, wine, bread and cheese, or the practice of selective breeding by which, over a period of 11,000 years, farmers have developed new breeds of animals and plants. Traditional breeding never entailed interference with DNA. No foreign DNA has ever been added or taken away in traditional biotechnology. Michael Pollan is in no doubt that Monsanto's genetically engineered potato 'New Leafs' is not in continuity with traditional breeding.

> Although Monsanto likes to depict biotechnology as just another in an ancient line of human modifications of nature going back to fermentation, in fact genetic engineering overthrows the old rules governing the relationship of nature and culture in a plant. For the first time breeders can bring qualities from anywhere in nature into the genome of a plant – from flounders (frost tolerance), from viruses (disease resistance) and in the case of my potatoes, from bacillus thuringiensis the soil bacterium that produces the organic insecticide known as Bt. The introduction into a plant of genes transported not only across species but whole phyla [1] means that the wall of that plant's identity – its irreducible wildness, you might say – have been breached.[2]

Recombinant DNA technology, on the other hand, cannot be called working with nature in any meaningful sense of that term. It is not natural for one species to cross-breed with a completely unrelated species, or for genetic material to be exchanged between unrelated species. Genetic engineering circumvents the barrier that exists between different species. It allows for the addition or deletion of proteins in ways not possible through reproduction, creating organisms that are

---

1. *Phyla* is the plural of *phylum* – a major grouping within the division of plants and animals that contain one or more classes. For example, the phylum Arthropoda contains insects, crustaceans, arachnids.
2. Michael Pollan, 'Playing God in the Garden', *New York Times* (Supplement), October 25, 1998. I am quoting from a text that was downloaded by the Irish environmental organisation VOICE, p. 2.

missing essential proteins or harbouring entirely new ones. Traditional forms of biotechnology leave the natural balance of genes, species and ecosystems intact. Genetic engineering effectively places control of evolution itself in the hands of molecular biologists.

Nevertheless we will see in the next chapter that the regulatory body for food in the US, the Food and Drug Administration (FDA), has chosen to treat transferred genes as natural food products as long as they come from an approved food source, thereby failing to consider the unpredictable effects which the old gene may have in its new system. Secret memoranda obtained from the FDA reveal that the agency ignored warnings from its own senior scientists who pointed out that genetic engineering is a new departure in plant biology and that it introduces new risks. It has emerged that the first GM crop to be commercialized – the Flavr Savr tomato – did not pass the required toxicological tests.[3]

Professor Bevan, a molecular geneticist and current Chair of the Working Group on Novel Foods in the European Union's Scientific Committee on Foods, disputes whether GE foods are safe for human consumption. He points to the unforeseen effects inherent in the technology. He is particularly worried about the next generation of GE foods, including the 'golden' rice or vitamin A-'enriched' rice. The reason these constitute a greater threat is because of the increased complexity of the gene construct.[4]

One does not need to be a Luddite to have serious questions about particular aspects of modern technology. The sad irony is that modern technology, especially in its chemical and nuclear phase, while it has delivered benefits to a segment of humanity in the form of better nutrition and longer lifespan, has wreaked havoc on the planet. Today, its impact is so

---

3. 'Open Letter from World Scientists to All Governments Concerning Genetically Modified Organisms', *Institute of Science in Society*, www.i-sis.org.uk/ 19/05/03, p. 5, No. 11.
4. Ibid. p. 5, No. 12.

extensive and damaging that it threatens the future of many life forms, including humankind itself.

There are many other examples where new technologies appear to benefit humankind on one level but, on the macro-scale, cause massive problems. The internal combustion engine brought increased mobility to many people in the twentieth century. Yet the increased number of cars around the world emitting greenhouse gases threatens to destabilise the planet's climate system. A few hundred million cars on good roads in different parts of the world will bring ease of mobility and comfort to many. Five billion cars, on the other hand, could have catastrophic consequences for humans and other creatures. What this means is that we need to look at the scale of our technology. This will inevitably mean that we will have to radically restructure much of our contemporary technologies for the simple reason that they are unsustainable in our finite world.

It is also true that in recent times many discoveries have been found to have very adverse, unintended effects which were not known when they were originally developed. There is increasing evidence that many of the 100,000 new chemicals which humans have invented over the past fifty years to improve the quality of human life are having a devastating impact on animal and plant life. In 1930 we produced one million tonnes of human-created chemicals. By 2000 the figure had jumped to 400 million tonnes.[5]

The widespread use of DDT began after the Swiss chemist Paul Muller discovered that it was a potent pesticide. In the 1940s it was used all over the world, especially in malaria eradication schemes in southern Europe, Africa, Latin America and Asia. DDT was cheap and it appeared harmless to everything except insects. Now we know that DDT is highly persistent in the environment and, like other organochlorines, it bioaccumulates in the fatty tissues of animals. In 1962, Rachel

---

5. Alok Jha, 'Tales from a Poisoned Planet', *The Guardian (Supplement)*, May 29, 2003, p. 6.

Carson's famous book *Silent Spring* detailed the impact of DDT on the reproductive behaviour of birds and in the wider environment. She pointed out, among other things, that DDT was linked with the thinning of egg-shells in many species of birds and their consequent failure to reproduce. Birds of prey like the eagle, osprey and peregrine falcon were particularly affected. It was also discovered that DDT caused cancer. In 1971, after a protracted battle against chemical companies that produced DDT, it was banned in the USA. Since then it has been banned in many Northern countries, though it is still used widely in the Third World.

For almost fifty years, chlorofluorocarbon (CFC) was considered to be the perfect chemical for refrigeration. Then in the 1970s it was discovered that CFCs destroy the ozone layer of the atmosphere which protects humans, other creatures and plants from the damaging ultraviolet rays of the sun. Many of the chemical companies that manufactured CFCs denied the connection until the mid-1980s. Eugene Linden claims: 'in the United States those who had the power to take action engaged in self-delusion: The Reagan Administration at first dismissed the ozone threat as a non-issue, while Du Pont and other manufacturers underestimated future sales of CFCs, making the hazard seem minimal.' [6] Both industry and the regulatory agency were very much to blame for not protecting the environment and the health of human beings.

While all the technologies mentioned have changed the human condition and the environment in various ways, none is as intrusive as genetic engineering, for the simple reason that it allows humans to scramble and re-programme the genetic code of all life forms on earth, including humans. This awesome possibility needs to be approached slowly and with great care. Commercial considerations must not be allowed to promote this technology without a long and thorough public

6. Eugene Linden, 'How the World Waited too long to Rescue the Shield That Protects Earth from the Sun's Dangerous UV Rays,' *Time*, May 10, 1993, pp. 56-8.

debate about the potential benefits to humankind and the earth and the possible nightmares that it might create.

WILL BIOTECH AGRICULTURE FEED THE WORLD?

The most common argument from proponents of this technology is that genetically engineered food and medicine will be necessary to feed a growing world population and to cure its diseases. They argue that if population levels rise to 10 billion it will be necessary either to increase land areas now under cultivation or to increase crop yields by new technologies like genetic engineering. They point out that it is almost impossible to extend farming any further because, where land is marginal, cultivating it will only exacerbate soil erosion.

In 1992, Monsanto's chief executive, Robert Shapiro, spoke along these lines in a long interview with Joan Magretta in the *Harvard Business Review*.[7] He argued that genetic engineering of food crops is a further improvement on the Green Revolution that saved Asia from starvation in the 1960s and 1970s. Similar arguments have been put forward by scientists, including Christopher Leaver, Professor of Plant Sciences at the University of Oxford. He points to the harsh realities of global population increase and diminishing agricultural lands. He claims that the only way to feed this growing population is through the use of gene technology. He also believes that it will be more environmentally friendly as it will involve the use of fewer chemicals in agriculture.[8]

Critics of genetic engineering reject the argument that GE foods will stave off global famine. They also question the accepted wisdom that the impact of the Green Revolution has been entirely positive. Dr Vandana Shiva, in correspondence with Norman Borlaug, considered by many to be the father of the Green Revolution, debunks many of the myths surround-

---

7. Robert Shapiro, 'Growth through Global Sustainability', *Harvard Business Review*, March 2, 1992, pp. 79-88.
8. Christopher Leaver, 'Novel Ways to Feed the World', *The Guardian*, February 17,1999, p. 8.

ing it. Dr Shiva challenges, first, the myth that India was unable to feed itself until the Green Revolution was launched. She points out that the last famine in India took place in 1942 during British rule. She admits that India experienced a severe drought in 1966 and was forced to import 10 thousand tons of grain from the US. She indicts the US administration who 'exploited this scarcity in its use of food as a weapon and forced non-sustainable, resource-inefficient, capital-intensive and chemical-intensive agriculture on one of the most ancient agricultural civilisations in the world. American agricultural experts like Borlaug did not introduce the Green Revolution to "buy time" for India. They introduced it to sell chemicals to India.' [9]

It is also important to remember that the Green Revolution is not simply a scientific story about hybrid crops, irrigation systems, cheap nitrogen and pesticides. John H. Perkins in his book *Geopolitics and the Green Revolution* [10] recounts the environmentally destructive and socially unjust aspect of the Green Revolution. In detailed case studies, Perkins insists that much of the energy behind the development of new varieties of crops stemmed from national security concerns in the US, Mexico, India, the United Kingdom and other countries. The theory was that, unless a growing population was able to harvest more food, there could be major security problems, especially the rise of Marxist guerrilla movements. This led planners to focus exclusively on increasing crop yield even at the expense of exacerbating social inequity and undermining biodiversity. Curbing population growth was also a key US foreign aid objective in the 1960s and 1970s.

It is also interesting to see how foundations closely connected to transnational corporations were instrumental in promoting the Green Revolution. These include both the Ford Foundation and the Rockefeller Foundation. It is worth

---

9. *The Ecologist,* Vol 27, No 5, September/October 1997, pp. 211-2.
10. John H. Perkins, *Geopolitics and the Green Revolution: Wheat, Genes and the Cold War,* Oxford University Press, New York, 1997.

remembering that the Green Revolution has contributed to the 'loss of three-quarters of the genetic diversity of major food crops and that the rate of erosion continues at close to 2 per cent per annum'.[11]

The same 'feed the world' arguments are being recycled by the promoters of genetic engineering today. In reality, famine and hunger around the world have more to do with the absence of land reform, with social inequality and biases against women in many cultures, and with a lack of access to cheap credit and basic technologies, rather than a lack of agribusiness super seeds. As we will see in the case of Argentina, planting GE soybeans, which is a cash crop, can even exacerbate poverty, especially for poor farmers.

This fact was recognized by the participants in the World Food Summit in Rome in November 1996. They acknowledged that the main causes of hunger are economic and social. People are hungry because they do not have access to food production processes or the money to buy food. Those who wish to banish hunger should address the social and economic inequalities that create poverty and not pretend that a 'magic' technology will solve all the problems.

My experience confirms this approach. I lived in Mindanao in the Philippines during the El Nino-induced drought of 1983. There was a severe food shortage among the tribal people in the highlands. The drought destroyed their cereal crops and they could no longer get food in the tropical forest because it had been cleared during previous decades. Even during the height of the drought, agribusiness corporations were exporting tropical fruit from the lowlands. There was sufficient rice and corn in the lowlands but the tribal people did not have the money to buy it. Had it not been for food-aid from NGOs, many would have starved.

Devindar Sharma, chairman of the Forum for Biotechnol-

---

11. Pat Roy Mooney, 'First Parts: Putting the Particulars Together', *Development Dialogue*, April 1998, p. 70.

ogy and Food Security in New Delhi, dismisses the claim that biotech food will feed the world. He told a meeting of the World Conference on Food and Farming in London in November 2002 that 'claiming that bio-technology or free-trade is needed to solve the problem [of hunger] is a deliberate distortion'.[12] He asked the audience to look at the experience of India under pressure from the WTO. 320 million people in India go to bed hungry each night despite the fact that India has 65,000 tonnes of food in storage. India is also exporting food because the poor cannot afford to buy the food. The government of India supports this policy because food exports bring in much needed foreign currency. But it does not help India's poor.

I find it interesting that Professor Leaver is silent about the economic and social factors, like land ownership, that gives rise to poverty and malnutrition. He confines his suggestions to hi-tech solutions, which, in my experience and that of others on the ground, usually benefit the better-off farmers. GE soya in Argentina has helped only wealthy farmers. Does Professor Leaver think that agribusiness companies will distribute genetically engineered food free to the hungry poor who have no money? Are land reform and economic policies aimed at helping small, subsistence farmers, no longer important? Is he not worried that genetic engineering will give enormous control of the staple foods of the world to a handful of Northern agribusiness companies? Most other people consider these companies to be dedicated, first and foremost, to making profits.

TERMINATOR GENE

The development by a Monsanto-owned company of what is benignly called a Technology Protection System, but what is more aptly called 'terminator' technology, is another rea-

---

12. Paul Brown, 'Hi-tech Crops Will Not Save the Poor', *The Guardian*, November 25, 2002, p. 9.

son for asserting that the 'feed the world' argument is spurious.[13] Because terminator seeds self-destruct after the first crop, this technology, if it becomes widespread, will surely strike the death knell for the 2.4 billion small, subsistence farmers who live mainly in the Third World. Sharing seeds among farmers has been at the very heart of subsistence farming since the domestication of staple food crops 11 thousand years ago. Terminator technology would effectively stop farmers sharing seeds. Hope Shand, research director with the Canadian ECT Civil Society Organisation (CSO) is alarmed at such a development.[14] 'Half the world's farmers are poor. They provide food for more than a billion people but they can't afford to buy seeds every growing season. Seed collection is vital for them.'[15] Terminator technology will enable Monsanto to control and profit from farmers in every corner of the globe. It will lock farmers into a regime of buying genetically engineered seeds that are herbicide-tolerant and insect-resistant, copper-fastening farmers to the chemical tread-mill.

For poor farmers in Third World countries, and the communities who depend on the food they produce, the widespread dissemination of terminator seeds will mean hunger, starvation and death. It is worth noting that the farmers of the South are the target market for terminator seeds. Delta and Pine have specifically suggested that rice and wheat farmers in countries like India, China and Pakistan are a priority market.

At an ethical level I suggest that a technology that, according to Professor Richard Lewontin of Harvard University, 'introduces a "killer" transgene that prevents the germ of the harvested grain from developing' must be considered grossly

13. Wayne Brittenden, '"Terminator" Seeds Threaten a Barren Future for Farmers', *The Independent*, March 22, 1998, p. 3.
14. ECT was formerly called RAFI (Rural Advancement Foundation International) is a civil society organisation devoted to Third World issues, especially those that deal with rural agriculture.
15. Quoted in John Vidal, 'Mr. Terminator Ploughs in', *The Guardian*, April 14, 1998.

immoral.[16] It is a sin against the poor, against nature and the God of creativity. Furthermore, if anything goes wrong, the terminator genes could spread to neighbouring crops and cause wild and weedy relatives of the plant that has been engineered to commit suicide. This would jeopardize the food security of many poor people. No wonder that there are those who consider it a form of biological warfare on subsistence farmers.

Some of the agribusiness companies promote their technology by talking about transferring the technology to the South. To date there has been very little transfer of genetic engineering technology from the transnational corporations to Third World countries. The appearance of 'terminator' technology shows that Northern companies are doing everything possible to avoid any such transfer. The World Bank Panel on Transgenic Crops concluded that technology transfer between transnational corporations and less developed countries was so rare that the examples they cited were exceptional.[17]

GENETIC ENGINEERING PROMOTED BY 'SPIN'

Given the huge financial stakes involved in biotech crops, it is understandable that all the stops are being pulled out in this battle for control of food production. The biotechnology industry has retained the services of a global PR company, Burson Marsteller, which specializes in crisis management and handling difficult or unsavoury situations. For example, it advised Babcock and Wilcox, the builders of the Three Mile Island nuclear installation in the US during the crisis in 1979. It also helped Union Carbide manage publicity in the after-

16. Jean-Pierre Berlan and Richard C. Lewontin, 'It's Business as usual', *The Guardian*, February 22, 1999, p. 14.
17. H. W. Kendall, R.. Beachy, T. Eisner, F. Could, F. Herdt, R.Vaven. P.H., J.S. Schell, and M.S. Swaminathan, *Bioengineering of crops: report of the World Bank Panel on Transgenic Crops*. International Bank for Reconstruction and Development/World Bank, 1997, Washington D.C.

math of the Bhopal tragedy in India which killed over 1,500 people. Among its clients, in the past few decades, were the repressive regimes in Indonesia, Argentina and South Korea.

In a document that was leaked to the press in August 1997 Burston Marsteller advised the biotech companies that 'they cannot hope to win the argument over the risks posed by genetically modified food, including the environmental dangers.'[18] The biotech companies were advised to focus on 'symbols, not logic'. These symbols ought to elicit 'hope, satisfaction, caring and self-esteem'.[19]

GE FOODS POSE A DANGER TO HUMAN HEALTH

Many people feel that the dangers posed by the tidal wave of biotechnological products are real and that a worldwide moratorium on the deliberate release of genetically engineered organisms must be put in place until the technology is much safer. At a meeting in Asilomar in the US in 1975, a group of scientists drawn from the Committee on Recombinant DNA of the US National Academy of Sciences, which included the Nobel Prize winner, James Watson, warned about the dangers of genetic engineering. They stated that 'there is serious concern that some of these artificially recombinant DNA molecules could prove biologically hazardous'.[20] This conference upheld the moratorium on recombinant DNA experiments. Jeremy Rifkin believes that the reason for the moratorium had more to do with the potential legal liabilities of creating bio-hazards than concern for human health or the environmental risks of the new technology.[21]

Almost twenty years later, an international group of scientists meeting in Malaysia in July 1994 called attention to the

18. Denny Penman, 'Stay Quiet on Risks of Gene-altered Food, Industry Told', *The Guardian*, August 6, 1997.
19. George Monbiot, 'The Fake Persuaders', *The Guardian*, May 14, 2002, p. 15.
20. Editorial, 'The Need to Regulate and Control Genetic Engineering,' *Third World Resurgence* No 53/54: 17.
21. Jeremy Rifkin, *The Biotech Century*, London, 1998, Victor Gollanz, p. xi.

scientific flaws inherent in the genetic engineering paradigm. They believe that genetic engineering is based on the false premise that each individual feature of an organism is encoded in one or more specific, stable genes and that the transfer of these genes results in the transfer of these discrete features. The truth is that no gene works in isolation but as part of an extremely complex genetic network. In fact, the function of each gene is dependent on the context of all the other genes in the genome. The same gene, for example, will have very different effects from individual to individual, because other genes in the genome are different.

The scientists who met in Malaysia pointed out that the development of any trait arises from many complex interactions between genes and their cellular context, and with the external environment. There are numerous layers of feedback mechanisms linking all these levels. These scientists insist that, in a significant number of cases, it is impossible to predict the consequences of transferring a gene from one type of organism to another. Furthermore, genetically engineered organisms, especially micro-organisms, may migrate, mutate and be transferred to other organisms and species. In some cases, the stability of organisms and ecosystems could be affected and threatened.[22]

Some of the risks to human health and the environment include the potential to cause allergies; an increase in antibiotic resistance and toxicity; misleading the consumer into thinking that the produce is fresh, and, finally, unpredictable gene expression in the engineered organism. Because of these risks, a group of scientists in the United States, calling themselves the Council for Responsible Genetics, have called for a more proactive approach from the regulatory agency which is the Food and Drug Administration (FDA) in monitoring and regulating genetically engineered foods.

22. Editorial, 'Some Critical Environmental Issues after Rio,' *Third World Resurgence* No81/82: 19.

Allergies

It is well-known that allergies in humans are caused by particular proteins. Genetic engineering involves adding new proteins to altered products. The FDA warns that new proteins in foods might cause allergic reactions in some people. People with food sensitivities and intolerances could be at risk from genetically engineered food. Because of lack of adequate labelling, these people may not be able to avoid such foods in the future. It is possible, for example, to transfer the gene for one of the many allergenic proteins found in milk into vegetables, like carrots. People who ought to avoid milk might not be aware that the transgenic carrots they might buy contained milk proteins.

It is important to emphasise that this problem is unique to genetic engineering. Genetic engineering routinely moves proteins into the food supply from organisms that have never been consumed as food by human beings. Some of those proteins could be food allergens, since virtually all known food allergens are proteins.

Recent research should alert the public to the possibility of an increase in allergenicity as a result of genetic engineering. A study by scientists at the University of Nebraska found that soybeans genetically engineered to contain Brazil-nut proteins caused reactions in individuals allergic to Brazil nuts. Blood serum from people known to be allergic to Brazil nuts was tested for the appropriate antibody response to the gene transferred to the soya bean. When seven out of nine volunteers responded to the genetically engineered soybeans, the researchers concluded that the allergenicity had been transferred with the transferred gene. Someone who was allergic to peanuts and who ate a food that contained peanut protein, could suffer a fatal allergic reaction.[23]

Scientists have a limited ability to predict whether a particu-

---

23. Celia Deane-Drummond, *Theology and Biotechnology*, Geoffrey Chapman, London, p. 86.

lar protein will be a food allergen if consumed by humans. The only sure way to determine whether a protein will be an allergen is through experience. Thus, importing proteins, particularly from non-food sources, is always a gamble from the point of view of allergenicity. It is generally recognized that there has been a significant rise in allergies, especially among children, in recent decades. With 8 percent of children showing allergic reactions to many commonly eaten foods, it seems foolish in the extreme to do anything that might increase allergenicity.

Antibiotic Resistance

Genetic engineering often uses genes for antibiotic resistance as 'selectable' markers. As we saw in Chapter Three, antibiotic resistant markers are used to determine which cells have taken up the foreign genes. Although they have no further use, the genes continue to be expressed in plant tissues. Most genetically engineered plant foods carry fully functioning antibiotic resistant genes. The most commonly used marker genes are the *npt11* gene that confers resistance to kanamycin, neomycin and geneticin, and the *bla* gene that confers resistance to ampicillin.

The presence of antibiotic resistant genes in foods could have two harmful effects. Firstly, eating these foods could reduce the effectiveness of antibiotics that are taken with such a meal. Antibiotic resistant genes produce enzymes that can degrade antibiotics. If a tomato with an antibiotic resistant gene is eaten at the same time as an antibiotic, it could destroy the antibiotic in the stomach.

Secondly, the resistant genes could be transferred to human or animal pathogens, making them impervious to antibiotics. If transfer were to occur, it could aggravate the already serious health problem of antibiotic-resistant disease organisms. It was claimed that unmediated transfers of genetic material from plants to bacteria is highly unlikely. But even a slight risk of this happening ought to require careful scrutiny in light of the seriousness of antibiotic resistance in the population at large. However, antibiotic resistant marker genes from GE plants have been found to transfer horizontally to soil bacteria and fungi in the laboratory.[24] Field tests revealed that GE sugar beet DNA persisted in the soil for up to two years after the GE crop was planted.[25]

In 1998, scientists at Cologne University discovered that DNA, which had been fed to a mouse, survived in the digestive system and subsequently invaded other cells in the mouse's body. The possibility for naked or free DNA to be taken up by mammalian cells is explicitly raised in the draft guidelines of the US Food and Drug Administration (FDA) with regard to antibiotic resistant marker genes. In 2000, reports on field experiments by the University of Jena in Germany indicated that GE genes may have transferred via GE pollen to the bacteria and yeast in the gut of bee larvae.[26] All of this recent scientific information should put a brake on the entry of GE food products into the food chain until there is much more research on their long-term impact on human health.

In July 2002, new evidence from British scientists working at the University of Newcastle raised further serious questions about the safety of GE foods. Research commissioned by the United Kingdom's Food Standards Agency (FSA) found that genes from antibiotic-resistant gene markers had found their way into the human gut. The scientists took seven volunteers who had had their lower intestine removed and were now using colostomy bags. The volunteers were given a burger with GE soya and a milkshake. The researchers compared their stools with twelve people with normal stomachs. The researchers found, to their surprise, that a relatively large proportion of genetically engineered DNA survived the pas-

24. Open Letter from World Scientists to All Governments Concerning Genetically Modified Organism, Institute of Science in Society, op.cit., p. 6, No. 22.
25. Ibid.
26. Ibid, p. 6, No. 19. See also A.Barnett, 'GM Genes "Jump Species Barriers",' *The Observer*, May 28, 2000.

sage through the small bowel. The research showed that no GE material survived the passage through the entire human digestive tract. Nevertheless, the very fact that antibiotic-resistant marker genes were identified for the first time in the human gut, something the proponents of GE food said could not happen, has given rise to a genuine fear that this could compromise antibiotic resistance in the population at large.[27] The fact that gut bacteria had taken up transgenic DNA is a cause for concern.

It is also true that the highly mosaic character of most vector constructs (used to bring about the transfer of genes) makes them structurally unstable and prone to recombination. According to scientists opposed to genetic engineering, this may be why viral-resistant transgenic plants generate recombinant viruses more readily than non-trangenic plants.[28]

Many organisms have the ability to produce toxic substances. These substances help the organism defend itself against predators in its environment. In some cases, plants contain inactive pathways leading to toxic substances. The addition of new genetic material, through genetic engineering, could reactivate these pathways. Alternatively it could increase the levels of toxic substances within the plants. This could happen, for example, if the on/off signals associated with the introduced gene were located on the genome in places where they could turn on the previously inactive genes. In the light of these considerations, many argue that genetically engineered foods pose new and unique challenges in terms of food safety.

Concentration of Toxic Metals

Some of the new genes being added to crops can remove heavy metals like mercury from the soil and concentrate them

---

27. Seán Poulter, 'Can GM Food Make Your Body Immune to Antibiotics?' *The Daily Mail*, July 17, 2002.
28. Mae Wan-Ho, Harmut Meyer and Joe Cummins, 'The Biotechnology Bubble,' *The Ecologist* May/June 1998, p. 149.

in the plant tissue. The purpose of creating such crops is to make possible the use of municipal sludge as fertiliser. Sludge contains useful plant nutrients, but often cannot be used as fertiliser because it is contaminated with toxic heavy metals. The idea is to engineer plants to remove and sequester those metals in inedible parts of plants. In a tomato, for example, the metals would be concentrated in the roots; in potatoes in the leaves. Doing this requires the use of genetic on/off switches that turn on only in specific tissues, like leaves.

Such products pose risks of contaminating foods with high levels of toxic metals if the on/off switches are not completely turned off in edible tissues.

There are also environmental risks associated with the handling and disposal of the metal-contaminated parts of plants after harvesting. This is a classic example of a technological fix for an environmental problem that ought to be addressed at its source. The way to guarantee that sewage sludge can be used in agriculture is to ensure that toxic substances do not enter sewage plants in the first place.

Generally, when people focus on health hazards associated with genetic engineering they concentrate on the genetic material that is added to organisms. There is also the possibility that the removal of genes can cause problems. For example, genetic engineering might be used to produce decaffeinated coffee beans by deleting or turning off genes associated with the production of caffeine. But caffeine helps protect coffee beans against fungi. Beans that are unable to produce caffeine might be coated with fungi, which can produce toxins. Fungal toxins, such as aflatoxin, are also toxic for humans and can remain active through the processes of food preparation.

Diminished Nutritional Quality

A possible consequence of genetically engineering foods is an alteration of the nutritional content of the resulting product. The FDA cautions that nutritional value could be significantly decreased without the crop exhibiting any outward signs. Humans have come to rely on certain characteristics of fruits and vegetables to indicate nutritional quality and flavour. For example, bright colour in peppers, apples and other fruits is generally associated with taste and ripeness. Genetic engineering may mislead consumers into buying fruits and vegetables which appear to be fresh and just about ripe, but, in fact, are engineered to last longer on the shelf and, as a consequence, may lack nutritional quality. This would have serious implications for public health and it needs to be taken on board by monitoring and regulatory agencies.

Potential Future Problems

History has shown that it takes a few decades for the full set of risks associated with any technology to be identified. In the 1920s, no one predicted that CFCs could cause such harm to the ozone layer. The ability to imagine what might go wrong with genetic engineering is limited by the current knowledge in such disciplines as physiology, genetics and nutrition.

We should not forget too quickly how pressures from the food industry and government agencies led to the failure of the UK authorities to link BSE in cattle with a new variant of the incurable human condition CJD. Those who raised questions about this connection in the mid-1980s were often criticised and even ridiculed by their colleagues. Dr Tim Holt, a Yorkshire doctor, told a UK government enquiry into BSE how a pathologist at the Government's central veterinary laboratory investigating Mad-Cow Disease, said that the transmission to humans was as 'unlikely as being struck by lightning'. [29]

The need for caution is highlighted by the controversy surrounding the production of transgenetic pigs to provide organs for human transplant operations. Companies on both sides of the Atlantic have engineered pigs to carry human

29. James Meike, 'BSE Warning That Was Ignored', *The Guardian*, 1 April, 1998, p. 1.

protein on the surface of their cells so that the organs will not be rejected by the human immune system. At first glance, this seems to be a practical way of meeting the demand for organs for transplant operations. Unfortunately, researchers have found that the pigs can carry at least two retroviruses. One of these has the potential to infect human cells. Even though the FDA have been provided with the results of the research, they have continued to allow the transplants to take place. One of the researchers involved felt that the least the FDA should have done is ban the transplants. [30] Given the presence of these viruses, many scientists would argue that pig organs can never be used safely in human transplants. [31]

GE FOODS POSE A DANGER TO THE ENVIRONMENT

Genes form a holistic system, with one gene affecting multiple traits and multiple genes affecting one trait. Consequently, scientists cannot always predict how a single gene will be expressed in a new system. For example, splicing a gene for human growth hormones into mice produces very large mice; splicing the same gene into pigs produces skinny, cross-eyed, arthritic animals. The FDA warns that splicing a single gene into an organism for a single desired effect may unintentionally cause other harmful reactions within that organism, which are not detectable.

Gene Transfer to Wild or Weedy Relatives

Organisms engineered to grow under adverse conditions run the risk of becoming weeds, either directly or by breeding with wild relatives. Here, weeds means all plants which are found in places where humans do not want to have them growing. In each case, the plants are found growing unaided and have unwanted effects as far as the farmer is concerned.

30. Phyllida Brown, 'Pig Transplants 'Should Be Banned",' *New Scientist* March 1, 1997, p. 6.
31. Michael Day, 'Tainted Transplants: Pig Organs May Never Be a Safe Replacement for Desperately Scarce Human Livers'. *The New Scientist*, March 1997, p. 6.

If GE plants become weeds, this could damage large tracts of agricultural lands, severely limit crop yields, and cause immeasurable destruction to sensitive ecosystems.

Some weeds result from the accidental introduction of alien plants, but many are the result of organisms introduced for agricultural and horticultural purposes. Johnson Grass that was intentionally introduced into the United States has become a serious weed. In Britain, serious damage is being caused by escaped mink that have no natural predator in that environment. In the Galapagos Islands, feral cats have wreaked havoc on the defenceless island fauna. In Africa, the South American Water Hyacinth has choked freshwater habitats. The ecologist Dr Bill Crowe points out that, in Ireland, introduced species like rhododendron and cherry laurel have clogged canals and choked shrub habitats. In Sligo, in 2001, whole bee populations had to be exterminated because of a disease introduced from Germany. In the past few years the zebra mussel has proliferated in the Shannon and is doing untold damage to pipes and other installations.[32]

It is completely plausible that a new combination of traits, produced as a result of genetic engineering, might enable crops to thrive in an environment in which they would then be considered a weed. For example, if a rice plant engineered to be salt-tolerant, escaped into a marine estuary, it could cause enormous damage. The possibility that this will happen increases as more and more genetically engineered organisms are released into the environment.

Biotech scientists and regulators often dismiss the possibility of genetically engineered crops becoming super weeds: they argue that most staple crops have been so weakened genetically by the domestication process that the addition of an engineered trait will not enhance their competitiveness. While this might be true of crops like corn, other crops –

---

32. Dr Bill Crowe 'GMOs in the Wild – What Are the Risks?', *Wild Ireland*, March-April 2002, pp. 46-8.

alfalfa, barley, potatoes, wheat, sorghum, broccoli, cabbage and radishes are examples – do retain their weedy traits. For example, a gene changing the oil composition of a crop might move into nearby weedy relatives in which the new oil composition would enable the seeds to survive the winter. The ability to survive winter cold might allow the plant to become a weed or might intensify the weedy properties it already possessed. This is why Dr Margaret Mellon, a molecular biologist, and Dr Jane Rissler, a plant pathologist, both of whom work with the Union of Concerned Scientists in the US, argue that the 'possibility that engineering will convert crops into new weeds is a major risk of genetic engineering.' [33]

Equally serious is the danger of what is called 'gene flow'. This refers to the possibility of transferring a gene from a transgenetic plant to a weedy relative by way of cross-pollination. Novel genes placed in crops will not necessarily stay in the fields in which they are planted. If relatives of the altered crops are growing near the field, the new gene can easily move, via pollen, into those plants. The new traits might confer onto the wild or weedy relatives the ability to thrive in unwanted places. If a herbicide resistant gene jumped to a wild weedy relative, that plant might become resistant to the particular herbicide. This form of genetic pollution could easily become a major nuisance to farmers worldwide.

European farmers fear that, if genetically engineered oilseed rape is planted in Europe, 'herbicide-tolerant rape will undoubtedly become part of the established volunteer weed populations that occur in many cereal rotations'.[34] The infestation may occur even where farmers do not grow the GE oilseed rape themselves. GE oilseed rape grown in an adjacent

33. Quoted in Rifkin, p. 87. Original reference: Rissler and Mellon, *The Ecological Risks of Engineering Crops*, 1996, MIT Press, Cambridge, MA, pp. 34-40.
34. G.R. Squire, D. Burn, and J.W. Crawford, 'A Model for the Impact of Herbicide Toleratnce on the Performance of Oilseed Rape as a Volunteer Weed', *Annals of Applied Biology*, 1997, pp. 315-338, quoted in *GeneWatch* May 1998.

field can pollinate plants in a neighbouring field and produce seeds that are herbicide tolerant. New research shows that cross fertilisation between GE crops and weeds can lead to a new generation of superweeds.[35] These will be resistant to the same herbicide that once made the crop special and wiped out their supposed advantage. Even worse, as we will see in Chapter Five, farmers whose lands are polluted by GE patented seeds can be prosecuted by the agribusiness company. In Canada, canola farmer, Percy Schmeiser, was found guilty of breaching patent rights and was fined when Monsanto's GE canola was found on his land.

Mistakes can also happen, as the following two examples demonstrate. In the late 1980s a company called Biotechnica International genetically engineered a micro-organism (Bradyrhizobium japonica) to improve nitrogen fixation in plants. The company contracted the Louisiana Agricultural Experiment Station to conduct field trials for one year by planting soybeans coated with the genetically engineered rhizobia. Rhizobia are rod-shaped bacteria which are found in the root modules of leguminous plants and are able to fix atmospheric nitrogen. After the experiment, the plants and seeds were incinerated, the fields were reploughed and replanted and Biotechnica ceased to have anything to do with the field trials. However, subsequent trials on that land revealed that the genetically engineered rhyzobia were outcompeting the indigenous strain. This was not expected to happen. The act of re-ploughing the area, rather than ending the experiment, in fact spread the genetically engineered rhyzobia over a four acre plot. The case illustrates the unpredictability of genetically engineered experiments according to one scientist:

> One of the major considerations about this case is that a microbe for which there existed an extensive historical database was used in a well-planned and thoroughly re-

35. 'Keep Britain GM Free', *The Ecologist*, July-August 2003, p. 37.

viewed experiment, and an unpredictable result was still obtained. [36]

A genetically engineered bacterium (Klebsiella) was found to produce dramatic changes in the soil food web and therefore to inhibit plant growth. The bacterium was engineered to produce ethanol from agricultural waste as a way of generating fuel. But when added to the soil it was found that it produced a significant decrease in growth in both roots and the shoots of wheat. There was also a decrease in beneficial soil fungi, an increase in parasitic nematodes (worms) and bacteria. Stopping the spread of such a bacterium, once released, will be very difficult. [37]

In November 2002, the US government ordered a company called Prodigene to destroy 500,000 bushels of soybeans that had been contaminated by genetically engineered maize. It appears that this company was attempting to grow different medicines, from Hepatitis B vaccine to an insulin-making enzyme, inside the kernel of genetically engineered corn. The company had planted a test field but, for reasons that are still unclear, the crop failed. Next, the company decided to plough up the field and plant it with conventional soya destined for the human food chain. Unfortunately, the gene from the medical crop was taken up into the soya. Thankfully, this was spotted before it reached the food chain.

It is estimated that experiments on GE pharmaceutical products are taking place at over 300 secret locations in the US alone. Some of these proteins are designed to act as vaccines or as contraceptives, to induce abortions, create blood clots, produce industrial enzymes, and propagate aller-

36. US National Biotechnology Impacts Assessment Programme Newsletter, March 1991. 'The Case of the Competitive Rhizobia'. Taken from the Greenpeace website. *www.greenpeace.org*
37. Mae Wan-Ho and B. Tappeser, 'Transgenic Transgression of Species' prepared for a workshop on transboundary movements of living modified organisms resulting from modern biotechnology, Aarhus, Denmark, 19-20 July 1996.

genic enzymes. One does not need to be a scientist to understand that, if any of these got into the food chain, the consequences would be truly terrible. [38]

In Chapter Three I wrote about the researchers who found that, despite the ban that was imposed by the Mexican government on Bt corn, genetically engineered maize contaminated local maize. Scientists are worried about this development as Mexico is the home of maize and acts as the gene bank for one of the world's staple crops. There are hundreds of varieties of maize in Mexico. The original report by two scientists from the University of California at Berkeley, Ignacio Chapela and David Quist, was carried in the November 2001 issue of the journal *Nature*. The March 2002 issue of *Nature* disowned Quist's and Chapela's research. However, in April 2002 the executive director of Mexico's National Commission on Biodiversity, Jorge Soberon, agreed that government tests had confirmed that the scale of the contamination was much worse than originally believed. A total of 1,876 seeds from two states, Oaxaca and Puebla, were examined, and Mexican scientists found evidence of contamination at 95 percent of the sites.

It is believed that grain that was meant for tortilla production was planted by Mexican farmers who were unaware that the corn was genetically engineered. Mexican scientists are convinced that the seeds came from either Monsanto, Aventis or Syngenta, as all three use the cauliflower mosaic virus as a promoter. However, Mexican scientists could not find out which of the companies was involved. The companies claimed commercial secrecy and refused to tell the Mexican authorities the relevant genetic information. Jorge Soberon found the company's behaviour irresponsible: 'I find it extremely difficult to accept this [the behaviour of the companies]. How can you monitor what is going on if they do not allow you the

---

38. 'Drop the Soy', *Guardian*, Society (Supplement), November 20, 2002, p. 8.

information to do it?' [39] Leaving aside the genetic disaster in this incident, the arrogance of the wealthy transnational corporation involved is mind-boggling. These companies are so powerful, financially and politically, that they feel no moral obligation to share their knowledge with officials of a sovereign government in order to track and prevent a worse ecological disaster.

It is also very worrying that the editor of the prestigious magazine *Nature* 'sided with a vociferous minority in obfuscating the reality of the contamination of one of the world's main crops with transgenic DNA of industrial origin.' [40] In a letter to the *The Guardian* the researcher in question, Ignacio H. Chapela, surmises: 'Perhaps the key lies in his tacit acknowledgment, albeit by dismissal, of the enormous pressures on anyone working in or around the biological sciences ever since we were set on a collision course with commercial interests.' He went on to sound a very chilling note about the future of science and the power of corporations to direct and manipulate research because they are funding it. 'The co-ordinated attempt to discredit our discoveries in the public piazza sends a chilling message to those who would dare ask important but uncomfortable questions and find their truthful answer. It is an assault on the very foundations of science.' [41]

According to the Mexican journalist, Tania Molina Ramirez, in October 2002 *Nature* refused to publish another report commissioned by the Mexican government, which confirmed the findings of Chapela and Quist.[42]

Herbicide Resistance

A second major environmental concern is the increased

---

39. Paul Brown, 'Mexicos' Vital Gene Reservoir Polluted by Modified Maize', *The Guardian*, April 19, 2002, p. 19.
40. Ignacio H Chapela, Letters column, *The Guardian*, May 24, 2002, p. 8.
41. Ibid.
42. Tania Molina Ramirez, 'Tainted Tortillas', *The New Internationalist*, January/February 2003, p. 19.

use of herbicides. Over half of the crops currently under development are being engineered for herbicide resistance, permitting increased use of these harmful chemicals. The permission granted in 1998 by the EPA in Ireland to Monsanto Ireland Ltd, to conduct field trials of sugar beet that has been genetically engineered to be resistant to the herbicide Roundup Ready raised questions about the nature of these herbicides and their increased use.

The company claims that even after long-term application there is no effect on the environment. Roundup Ready is marketed as an environmentally friendly herbicide. Monsanto also claims that the use of seeds genetically engineered to be resistant to Roundup Ready would lead to a decrease in the use of herbicides. Such claims need to be thoroughly scrutinized. Roundup Ready is a broad spectrum non-selective herbicide which kills all plants, including grasses, broadleaf trees and woody plants. The active agent in Roundup Ready, glyphosate, is an organophosphate. Unlike other organophosphates, it does not affect the nervous system of animals; but that does not make it environmentally friendly. The one claim you cannot make for glyphosate is that it is environmentally friendly.[43] Many species of wild plants are damaged or killed by applications of less than 10 micrograms per plant. These plants are particularly vulnerable when it is spread from the air. Fish and invertebrates are also very sensitive to formulations of glyphosate, as are beneficial insects like lacewings and ladybirds.

Critics of Monsanto point out that it is very difficult to measure glyphosate residue in the environment. Only a few laboratories have the sophisticated equipment and the necessary expertise. This means that data are often lacking on residue levels in food and in the environment, and existing

---

43. 'Open Letter from World Scientists to All Governments Concerning Genetically Modified Organisms', op.cit., p. 4, No. 9. See also H Hardell and M Erisson, 1999, 'A Case-Control Study of Non-Hodgkins Lymphoma and Exposure to Pesticides', *Cancer 85*, 1355-1360.

data may not be fully reliable.

While the acute toxicity of glyphosate for mammals is very low, it can interfere with some enzyme functions in animals. In California, glyphosate is the third most commonly reported cause of pesticide-related illness among agricultural workers. A 1999 study links glyphosate with non-Hodgkins lymphoma. More recent studies have shown that the problems with Roundup Ready stem not so much from the glyphosate as from the unlabelled 'inert' ingredients that aim to make the herbicide more efficient. According to Joseph Mendelson, Roundup Ready consists of 99.04 per cent 'inert' ingredients, many of which have been identified, including polyethoxylated tallowamine surfactant (known as POEA), related organic acids, flyphosate, isopropylamine, and water. Researchers have found that the acute lethal dose of POEA is less than one-third that of glyphosate alone. Studies by Japanese researchers on victims of poisoning discovered that this 'inert' ingredient increased toxic levels in patients.[44]

There is also evidence that plants that are grown in the presence of weed-killers can suffer from stress. They react by producing or failing to produce certain proteins or substances. Members of the bean family produce higher levels of plant-oestrogens (phyto-oestrogens) when grown in the presence of glyphosate. Excessive levels of these oestrogens present a risk to unborn babies and to children. These plant-oestrogens mimic the role of hormones in the human body. They can be particularly disruptive of the human reproductive system, especially for young males.

There is also good reason to be skeptical about claims that genetically engineered plants will lead to fewer chemicals in agriculture. In soybean cultivation Monsanto maintained, in documents prepared for the US authorities, that it now takes between one and five applications of different herbicides to

---

44. Joseph Mendelson, 'Round-up: The World's Biggest-Selling Herbicide,' *The Ecologist*, September/October 1998, p. 272.

control weeds and that, with Roundup Ready, only one or, possibly, two applications will be needed. Yet in their advice to farmers in Argentina, Monsanto recommended that Roundup Ready be used with the GE soybeans before sowing, when the plants are young, after three or four leaves have appeared, and whenever the farmer finds weeds. This is quite a different scenario.[45] A survey of 8,200 field trials of the most widely grown GE crop, herbicide-tolerant soya beans, found that the yield was 6.7 percent below the non-GE soya, and that the GE crop required two to five times more herbicides than the non-GE varieties.[46] This was further confirmed by other studies that found the GE crops performed erratically, were more susceptible to disease, and provided poorer economic return to farmers.[47]

In the light of research which showed that Roundup Ready caused human health problems and environmental damage, Monsanto agreed in January 1997 to change its advertisements for glyphosate-based products. This was in response to complaints by the New York Attorney-General's office that the advertisements were misleading. As part of the agreement, Monsanto will not use the terms 'biodegradable' or 'environmentally friendly' in its advertisements for glyphosate-based products in New York State. They also agreed to pay $50,000 towards the State's cost of pursuing the case. Monsanto claims it did not violate any federal, state or local law and that its claims were true and not misleading in any way. The company states that it entered into the agreement for settlement purposes only in order to avoid costly litigation.[48] In November 1997, the Dutch Advertisement Code Committee (ACC) found that Monsanto's advertisements for Roundup Ready were misleading. The ACC judged that Monsanto's herbicide

---

45. Greenpeace International Glyphosate Fact Sheet, November 1996.
46. 'Open Letter from the World Scientists to All Governments Concerning Genetically Engineered Crops', p. 3, No. 3.
47. Ibid.
48. Factsheet from Genetic Concern, *Round-Up!Round-Up*, April 1998.

is not biologically degradable and that their eco-friendly claim is in conflict with the truth.

Companies like Monsanto claim that by producing crops that are resistant to herbicides they will reduce the amount of harmful chemicals entering the environment. They fail to inform the public that it is also a very cost-effective operation for the company. At present, it costs between $40 and $100 million to bring a new pesticide through the regulatory process to farmers' fields. It only takes $1 million to develop a new plant variety. Pat Mooney believes that 'economics dictate that chemical companies invent new crop varieties adaptable to the company's chemicals rather than adapt expensive pesticides to the inexpensive seeds.' [49]

Problems for Biodiversity

We have already seen how agribusiness has contributed to the huge loss of biodiversity in the world in recent years, because it opts for monoculture hybrid seeds. Genetic engineering will exacerbate the threat to biodiversity. The UK Advisory Committee on Releases to the Environment is concerned that when crops which are genetically engineered to be resistant to herbicides become common they will have a devastating impact on wildlife. Fields of genetically engineered crops could lead to starvation for birds and insects that depend on seeds as a source of food. The committee's chairman, John Beringer, Professor of Molecular Genetics at Bristol University, said that 'it could be cranking up the pressure on species if this technology proceeds to the limits'.[50]

Vandana Shiva, an Indian scientist who studies biodiversity, claims that genetic engineering, even at present, is working against crop diversity and is narrowing the genetic base of agriculture to only a few crops. In 1998, two commercially

---

49. Pat Roy Mooney, 'Private Parts: Privatisation and the Life Industry' *Development Dialogue*, 1998, p. 147.
50. Nicholas Schoon, 'Genetic-crop Threat to Wildlife Survival,' *The Independent* March 25, 1998.

staple crops were being genetically engineered – soya and maize. These two crops are now destined to take the place of hundreds of legumes, beans and cereals like millet, wheat and rice. She goes on to point out that genetically engineered crops are based on expanding monocultures of the same variety evolved for a single function. Bill Crowe, University College, Cork, notes that these large companies develop only a few varieties of any seed and with machine handling in view. Small farmers use hand-dispersal methods for sowing seeds; where machines are used, the ear of the grain must be of a consistent size and texture. Such seeds invariably require high nutrient inputs and petrochemical herbicides that wipe out everything else and leave nothing for other species like birds. As the biotechnology industry takes root in different countries, this monoculture tendency will continue, further undermining agricultural biodiversity and, thereby creating ecological vulnerability.

Threatens Traditional Insecticides

Many insects have genes that render them susceptible to pesticides. Often these susceptible genes predominate in natural populations of insects. These genes are a valuable natural resource because they allow pesticides to remain as effective pest-control tools. The more benign the pesticide, the more valuable the genes that make pests susceptible to it.

Certain genetically engineered crops threaten the continued susceptibility of pests to one of nature's most valuable pesticides: the *Bacillus thuringiensis* or Bt toxin. These Bt crops are genetically engineered to contain a gene for the Bt toxin. Bt-resistant insects have already evolved because of their continual exposure to the toxin in Bt plants throughout the growing season. As a result, the US EPA has recommended that farmers plant up to 40 percent non-GE crops in order to create 'sanctuaries' for non-resistant insect pests.[51] In fact, in

---

51. 'Open Letter from World Scientists to All Governments Concerning Genetically Modified Organisms', op. cit., p. 4, No. 8.

both laboratory and field situations a number of species, including the Colorado potato beetle, have developed a resistance to the Bt toxin. In 1996 Monsanto's Nu Corn which contained the Bt toxin, failed to perform as expected due to hot weather and drought conditions. Even in field tests on Bt Cotton the genetically engineered gene killed only 80 percent of the boll-worms that attack cotton. The fact that 20 percent survived means that a 'super bug' resistant to Bt will inevitable emerge.

There is a culture of secrecy surrounding genetic engineering. One can add to that the commercial pressure which giant corporations can bring to bear on governments. So it comes as no surprise that an EU study which showed that genetically engineered crops would raise the costs of production for all farmers and threaten organic farmers, was kept under wraps for eighteen months. The study, undertaken by the Institute for Prospective Technological Studies, of the EU Joint Research Centre, addressed the question of whether it would be possible to have GE and non-GE agriculture coexisting. It found that the cost would be prohibitive and would place particular burdens on organic farmers. The most worrying aspect is that the study was delivered to the EU Commission in January 2002 with the recommendation that it not be made public. Luckily, it did reach the public when someone leaked the document to Greenpeace in May 2002.

Groups like Greenpeace and Friends of the Earth, rather than governments, have taken on the responsibility of providing the public with data on the safety of their food. Most governments have promoted the agribusiness's GE campaigns vigorously. The corporations do not want people to be told what the genetic sources of their food is. In March 2002 a ruling by a quango of the Food and Agricultural Organisation (FAO), called the Codex Alimentarius, stated that governments cannot demand that consumers be told their food's genetic origins – whether a cooking oil comes from soya or corn that was genetically engineered. The only exception is when a food turns out to be dangerous by causing allergies or other reactions.[52] The committee did not seem to realise that *post factum* traceability is almost impossible. Given all the chemical elements now in the food chain, the corporations could easily point the finger at one of these instead of at their own products. This is another example of governments caving in to corporate demands even when a threat to human health is a possibility.

Poisoning Nature

The addition of foreign genes to plants could also have serious consequences for wildlife. For example, engineering crop plants, such as tobacco or rice, to produce plastics or pharmaceuticals could endanger mice or deer who consume crop debris left in the fields after harvesting. Fish that have been engineered to contain metal-sequestering proteins (such fish have been suggested as living pollution clean-up devices) could be harmful if consumed by other fish or raccoons.

Creating New, More Virulent Viruses

One of the most common applications of genetic engineering is the production of virus-tolerant crops. Such crops are produced by engineering components of viruses into the plant genomes. For reasons not well understood, plants producing viral components on their own are resistant to subsequent infection by those viruses. Such plants, however, pose other risks, like creating new or worse viruses through recombination.

Recombination can occur between the plant-produced viral genes and closely related genes of incoming viruses. Such recombination may produce viruses that can infect a wider range of hosts or that may be more virulent than the parent viruses.

---

52. 'Eat What You're Given and No Arguments', *The New Scientist*, March 16, 2002, p. 12.

In the late 1980s, the National Institute of Allergy in the US sought an animal model suitable for studying AIDS. Researchers introduced the AIDS virus into mice. Critics of the experiment feared that if the AIDS-infected mice escaped this could create a new and even more deadly source of AIDS infection. Those conducting the experiment dismissed such fear as unfounded and alarmist. A study conducted by Dr Robert Gallo, one of the co-discoverers of the AIDS virus, and subsequently published in the magazine, *Science*, cautioned against using animal research models. He and his colleagues argued that the AIDS virus carried by the experimental mice might combine with other viruses carried by mice. This could result in the creation of a new more virulent form of AIDS that could be transmitted in novel ways, even through the air.[53]

Genetically Engineered Crops Could Devastate Third World Agriculture

There is the concern that genetically engineered crops will displace crops grown naturally by farmers in Third World countries and, in the process, disrupt the lives of millions of poor people. In the US, two biotechnology companies have produced vanilla from plant cell cultures in laboratories. The price of naturally produced vanilla is about $1,200 per pound. The biotechnology companies estimate that they can commercially produce genetically engineered vanilla for about $50 per pound. Such a development would wipe out the livelihood of about 100,000 farmers in Third World countries. Developments like these could constitute an economic disaster for many Southern countries where biodiversity is already under severe strain.

Similar research is under way to genetically engineer crops that are crucial to Third World economies. These include coffee, tobacco, cocoa, coconut, palm oil, sugar and ginseng. Genetically engineered varieties may thrive in temperate zones and thus ravage many Third World economies that are

53. Jeremy Rifkin, *Biotech Century*, p. 73.

dependent on one or other of these commodities. These countries have no fall-back industries capable of absorbing their redundant farmers. From living in the Philippines for many years, I know that the lives of millions of copra farmers would be devastated if coconut oil was produced in temperate zones.

Unforeseen Problems

As with human health risks, it is unlikely that all the potential risks to the environment have been identified. At this point, biology and ecology are too poorly understood to be certain that scientists can comprehensively rule out any major and irreversible damage to the environment. Therefore we ought to proceed with extreme caution. This point was made forcibly by Robin Grove-White, Director of the Study of Environmental Change, Lancaster University, in the wake of Prince Charles's article on genetic engineering which appeared in the *Daily Mail*.[54] Mr Grove-White wrote:

> Last year, in a study sponsored, to its credit, by Unilever (itself a potential beneficiary of the technology), we found that the panoply of ministerial advisory committees and other regulatory mechanisms is failing utterly to engage with issues of central significance to most people – particularly, the unknowns surrounding future cumulative dependency on genetically engineered crops and foods, with the risks of unforeseen (because unforeseeable in terms of current scientific understanding) synergies and ecological or public health mishaps.[55]

Concerns about gene pollution and the medical consequences of an accident are beginning to appear in modern fiction. In the novel *Carriers*, the author Patrick Lynch makes a very plausible case for believing that a disaster can follow from a genetic engineering experiment that went wrong even

---

54. 'My 10 Fears for GM Food', *Daily Mail*, June 1, 1999, pp. 10 and 11.
55. Robin Grove-White, Letters column, *Independent*, June 14, 1998.

though no one could imagine what the consequences of the gene therapy could be at the time.⁵⁶ In May 1999 the British Medical Association (BMA) called for an 'open-ended' moratorium on GE crops because of the doctors' concern about the potential health and environmental risks.⁵⁷

ARGENTINA 2002

As a case-study of what can happen, it is worth assessing the impact of GE soybeans on Argentina. When Monsanto brought the technology to that country in 1996, about 90 per cent of the soya farmers opted for GE soya. As a result, Argentina's soya crop has doubled to 27 million tonnes. That might seem like good news; but the social and environmental costs have been high. The growth in soya output has not come from higher yields from the GE soya but from more land devoted to soya. In fact, GE soya has had a 5 or 6 per cent lower yield than conventional soya. The promised reduction in herbicide use has not materialized. In fact, many farmers are using two to three times more herbicide than they used with the traditional crop. It is estimated that costs have risen by about 14 per cent; but since overproduction has caused the price to drop, the farmers are actually poorer.

The one efficiency that GE soya has brought is very worrying, ecologically and socially. Everyone will admit that cultivating GE soya saves time for the GE soya farmers since they do not have to perform the traditional tasks of ploughing and harrowing. What they do is to drench the land with herbicide (Roundup Ready) and then sow the soya seed directly on the land. Such methods facilitate the larger holdings and this, in turn, has put small farmers out of business. The impact on the people has been devastating, with knock-on damage to small market towns that depend for their livelihood on a thriving agricultural sector. Many people have been forced off the

56. Patrick Lynch, *Carriers*, Random House Publishing, 1997, New York.
57. Kevin O'Sullivan, 'BMA Wants Moratorium on GM Crops while Potential Risks are Examined', *The Irish Times*, May 18, 1999, p. 6.

land and have migrated to cities where they now live in squalor in slums.

The ecological consequences are even worse. Forests are being cleared to plant more GE soya to compensate for the fall in prices. At the moment, farmers are spreading 80 million litres of herbicide on 10 million hectares under GE soya. The chemicals kill everything except the GE soya crop. This is affecting the humus quality of the soil which no longer can retain moisture. Traditionally, farmers grew soya in the summer and wheat in the winter. But now this rotation no longer happens; there is nothing but soya. This is not sustainable in the long term. No wonder that, recently, one of Argentina's leading agronomists, Jorge Eduardo Fuli stated that 'our brief history of submission to the world bio-technology giants has been so disastrous that we fervently hope other Latin American nations will take it as a example of what not to do.' [58]

The pressure to promote transgenic food is coming not from the farmers or consumers but from the biotech corporations. Looking back over the last eight years from the vantage point of May 2003, it is clear that vast majority of genetic manipulation has been undertaken for commercial gain. Friends of biotech companies in the Clinton and Bush administrations have used every opportunity to promote GE food globally. Biotech companies believe that they can make huge profits if farmers and consumers are pressurized into planting and buying genetically altered food. This is simply greed and we should not allow spurious arguments to obfuscate that fact. What is at stake here is not some issue of minor importance to the earth and humankind. We are talking about manipulating the food sources of our world.

---

58. Sue Branford, 'Why Argentina Can't Feed Itself: How GM Soya Is Destroying Livelihoods and the Environment in Argentina', *The Ecologist*, October 2002, p. 23.

FIVE

An Unholy Trinity – Regulatory Agencies, Biotech Corporations, and Governments

APART FROM the serious dangers to human health and the environment which I outlined in Chapter Four, there are other reasons for demanding a moratorium on the deliberate release of genetically engineered (GE) organisms into the environment. These have to do with the poor regulatory regime that currently evaluates GE products, especially in the US, and the power of the biotech industry to influence political decisions.

As far back as 1997, the Council for Responsible Genetics in the United States urged the US Food and Drug Administration (FDA) to re-evaluate its position on genetically engineered foods. The essential elements in such a policy ought to include:
– adequate and independent testing;
– proper and complete registration of genetically engineered foods;
– separation at source of genetically engineered crops;
– clear and accurate labelling.

TESTING

Testing is crucial. The tests must be scientific and comprehensive. Unless you look for something special in your research questions, you will not find it. Industry and government spokespersons have been assuring the public that genetically engineered organisms are safe. They fail to inform the public that the testing regime is inadequate and often works from quite contradictory positions. The British government's Review Group on GM Food stated; 'To date there has been no verifiable untoward toxic or nutritional bad effects from GM food, but that does not mean there will not be any.'[1] There has been no study of the effects on people of eating GM food, and none has yet been developed.[2]

An article in the *New York Times*, entitled 'Playing God in the Garden', by Michael Pollan, illustrates how unsatisfactory the present regulatory regime is.[3] Pollan reminds his readers that they may be eating genetically engineered soya, corn or potatoes without knowing it. Even though genetically engineered foods have been on the market for five years in the US, the regulatory agency for food, the U.S. Food and Drug Administration (FDA), does not require that genetically engineered food be labelled as such. As we will see later, the situation in Europe is very different. Biotech companies lobbied the US government and the FDA to ensure that the industry would not be forced to inform the public that the food they were eating was genetically engineered.[4]

The author goes on to point out that a genetically engineered potato on sale as food, Monsanto's New Leaf Superior potato, is, itself, registered as a pesticide with the U.S. Environmental Protection Agency (EPA). This potato has been genetically engineered to poison and kill the Colorado potato beetle. Every cell of Monsanto's New Leaf Superior contains a gene from the *Bacillus Thuriengensis* bacterium (Bt.) which is highly toxic to Colorado potato beetles. This is why this potato is registered as a pesticide.[5]

1. 'Unexplored Threats to Health, Wildlife and Biodiversity', *The Guardian*, July 22, 2003, p. 9.
2. ibid.
3. Michael Pollan, 'Playing God in the Garden', *New York Times* (Supplement) October 25, 1998. I am using a print-out version which was circulated by the Irish environmental organisation VOICE,
4. Ibid p. 1.
5. Ibid.

While the FDA has responsibility for licensing food, the USEPA has responsibility for licensing new pesticides. According to Pollan, the EPA pesticide officials believe that the New Leaf Superior potato is reasonably safe for humans. In an experiment, EPA scientists fed pure Bt to mice without causing them harm. Because humans have eaten old-style New Leaf potatoes for a long time, and because mice are not visibly harmed by eating pure Bt, the EPA concluded that potatoes containing Bt genes are safe for humans.

Pollan reported that some geneticists believe this reasoning is flawed because, as we have seen earlier in this book, inserting foreign genes into plants may cause subtle changes that are difficult to recognize.[6] As we will see, the research of Dr Arpad Pusztai, at the Rowett Insitute in Scotland, raised serious concerns because when he fed Bt potatoes to rats many organs shrank and their immune system was affected.

When consumers go to the supermarket to buy a bag of Monsanto's New Leaf Superior potatoes they will find a list of all the nutrients and micro-nutrients in the potato printed on the bag. They will not learn that the potato has been genetically engineered or that it is legally a pesticide. The reason for this anomaly is a bureaucratic mismanagement between officials in two government agencies responsible for human and environmental welfare who do not seem to communicate with each other on a very basic question of human well-being.

In the USA, food labelling is ordinarily the responsibility of the FDA. An FDA official told Pollan that the FDA does not regulate Monsanto's potato because the FDA does not have the authority to regulate pesticides. According to them, that is EPA's job. The farce deepens when one realises that an EPA-approved pesticide would normally carry an EPA-approved warning marker. For example, a label on a bottle of Bt will warn the user not to inhale the substance or allow it to come in contact with an open wound.[7] However, in the case of

6. Ibid p. 4.
7. Ibid.

Monsanto's genetically engineered potato, with the Bt gene, the EPA insists that it is the responsibility of the FDA to label the item since the potato is a food and, therefore, comes under the remit of the FDA. However, an FDA spokesperson informed Pollan that it only requires genetically-engineered foods to be labeled if they contain allergens or have been 'materially changed'.[8] In the case of the genetically engineered potato the FDA has determined that Monsanto did not 'materially change' the New Leaf potato by turning it into a pesticide. Therefore, no FDA label is required.

Furthermore, the law that set up the FDA (the Food, Drug and Cosmetic Act) forbids the FDA from including information about pesticides on food labels. Pesticide labels are EPA's responsibility, according to the FDA. While two agencies quibble about who has responsibility for what, the consumer is faced with eating food that is potentially harmful. Neither agency will guarantee the safety of staple foods.

The corporation that produced the potato does not feel that food safety is its responsibility either. A Monsanto official told Pollan that the corporation should not have to take responsibility for the safety of its food products. Monsanto should not have to vouch for the safety of biotech food, according to Phil Angell, Monsanto's director of corporate communications. 'Our interest is in selling as much of it as possible. Assuring its safety is the FDA's job'.[9]

Apart from important decisions falling between various agencies, there is also the problem of government agencies being under-resourced and therefore not putting risk-assessment high on their agenda. In 1998 the US Department of Agriculture was spending only one percent of the funds allocated to biotechnology research on risk-assessment.[10]

During four years as a member of the Irish EPA Ethics Advisory Board, I learned that researchers have also chal-

---

8. Ibid. p. 5.
9. Ibid.
10. Jeremy Rifkin, *The Biotech Century*, Victor Gollancz, London, p. 77.

lenged the adequacy of the current field-testing procedures for GE crops. They argue that since the tests are designed to rule out 'gene flow' they are faulty. Field-testing procedures require early harvesting of the crop or, alternatively, culling the flowers on a crop like potatoes. Such flawed procedures cannot give adequate data to the regulatory agency to allow them to assess accurately whether there will be a major risk associated with a large scale commercial planting of a transgenic crop. Furthermore, the fact that the experimental area is small and the time scale is limited to one or, at the most, a few harvests means that there is little possibility of assessing the negative impact on micro-organisms, insects and plants over a longer period of time. Extrapolating to the wider environment inevitably brings considerable scientific uncertainty, given varying climates and agricultural practices.

Most trials are designed to evaluate agronomic characteristics (e.g. yield) rather than the the impact on health or the ecology. There needs to be rigorous testing for the nutritional value of the food and for the toxicological and allergenic implications of eating GE food.

Finally, current tests are carried out on a single GE crop. A single crop-testing regime is a very poor guide for judging the potential danger to the environment when two or twenty GE crops are planted in close proximity to each other.[11] Studies are currently conducted on a case-by-case basis, neglecting the potential for cumulative impacts as ever increasing numbers of herbicide-resistant crops are grown. With regard to human health, testing has, to date, relied on laboratory studies with laboratory species.[12]

Moreover, in their testing programmes biotech companies do not always comply with regulations when conducting trials in First World countries. In February 1999, Monsanto was fined £17,000 for failing to observe the six-feet buffer-zone at

11. Ibid. p. 78.
12. Jeremy Lennard, 'Washington Kills Global Pact to Govern GM Trade', *The Guardian*, 23 February, 1999, p. 14

a test site in Lincolnshire where GE oilseed rape was being grown. It also appeared that funding for monitoring was grossly inadequate. A mere £80,000 was allocated to monitor the 340 test sites in Britain during 1998. As a result, only 70 sites were visited. In August 2002, Aventis admitted that they had planted the wrong type of seed for three years in fourteen rape seed fields across the country as part of GE crop trials.[13]

PROBLEMS WITH REGULATORY AGENCIES

In the light of such examples, is it any wonder that many environmental and consumer groups are very unhappy with national and international regulatory agencies? In the US, as we have seen, the present system, in most cases, is that genetically engineered products do not require pre-market approval, public notification, or any labelling whatsoever to inform consumers of their novel, and possibly harmful, characteristics. The FDA does have the power to regulate food, but in the case of most genetically engineered foods has chosen not to do so. As we saw in Michael Pollan's article, the FDA maintains a list of foods that need no regulation because they are 'generally recognized as safe' (or GRAS).[14]

Since 1992, the FDA has allowed companies like Monsanto to decide for themselves whether their new genetically-engineered foods should be added to the GRAS list and thus escape regulation. In other words, FDA regulation of genetically engineered foods is voluntary, not mandatory. This is why the Council for Responsible Genetics claims that a precautionary, safety first, policy has been scrapped in favour of corporate economic interests.[15]

13. Seán Poulter, '£17,000 Fine That Will Take 90 Seconds to Pay', *Daily Mail*, February 18, 1999, p. 9.
14. Michael Pollan, art. cit., p. 5.
15. The Council for Responsible Genetics is a non-governmental organisation which seeks to foster debate about the social, ethical and environmental implications of genetic engineering. CRG works through the media and concerned citizens to distribute accurate information, and represents the public interest on emerging issues in biotechnology. Its website is www.gene-watch.org

Industry is essentially placed on an honour system, itself deciding when and whether to consult with the FDA. Companies conduct safety tests for their own bio-engineered products. They only have to notify the FDA if they suspect a problem. If they think there is no danger to consumers, companies are not required to state that their product has been genetically manipulated or to reveal the source of implanted genes. Neither are they required to make the results of their safety tests available to the public.

In this lax regulatory climate, if something goes wrong the FDA will not have a complete set of data regarding genetically engineered foods on the market. Thus, there will be no way to trace which GE food has caused the problem. At the moment, the FDA policy denies consumers' right to know how their food has been grown and processed. It also removes the basic human right to safe and tested food by allowing the companies who profit from biotechnology to decide if and when a product is hazardous.

More worrying still, consumer and environmental groups also claim that Monsanto, and other corporations, have successfully co-opted national and regulatory agencies to promote their agenda. The revolving door syndrome whereby high-ranking personnel from the corporate world move into critical positions in the FDA and then back to industry, raises questions about the thoroughness and impartiality of the FDA and other regulatory bodies. For example, Michael Taylor, the FDA official who wrote the guidelines which prohibit farmers or dairy companies from labelling their milk as free from the Monsanto bovine growth hormone (Bovine Somatotropin), spent seven years as a Monsanto corporate lawyer.[16] The reason why dairy farmers with huge herds inject their cows with the bovine growth hormone is to increase their milk output by 15 to 20 per cent and thereby increase their profits. Farmers' profit should not take precedence over human and

16. Daniel Jeffreys, 'The Record That Shames the Biotech Bullies', *The Mail*, February 18, 1999, p. 8.

animal health.

In 1994, an FDA 80-page study, *Use of Bovine Somatotropin in the United States*, concluded that there is no evidence that the GE bovine growth hormone poses any threat to human or animal health. British scientists, however, disagreed. Their data revealed that bovine growth hormone may increase a cow's susceptibility to mastitis. The British scientists showed that Monsanto's submission to the FDA was based on selected data and covered up what research had revealed, namely that there was more pus in the udder of cows that were treated with bovine growth hormone. A report by Canadian government scientists claims that the hormone harms cows, and states:

> Evidence from the animal safety reviews were [*sic*] not taken into consideration. These studies indicated numerous adverse effects in cows, including birth defects, reproductive disorders, higher incidence of mastitis [infection leading to inflammation of the udder], which may have had an impact on human health.[17]

Rather than face these problems, Monsanto chose to stop milk suppliers telling their consumers that there is a genetically engineered hormone in their milk. Food companies which marked their products as GE-free have been threatened and sued by Monsanto. For example, Monsanto brought an action against the Pure Milk and Ice Cream Company in Texas to try to force them to withdraw their labels.

George Monbiot, writing about the FDA's handling of Monsanto's GE bovine growth hormone controversy, observed: 'The administration has approved some of the company's most controversial products, including the artificial sweetener aspartame and the injectable growth hormone for cattle. Only the New York Attorney General's office has taken the company to task, forcing it to withdraw adverts claiming that

17. Shiv Copra and others, 1998, RBST (NUTRILAC) 'Gaps Analysis' Report by RBST Internal Review Team, Health Protection Branch, Health Canada, April 21, 1998. p. 29.

Roundup Ready is biodegradable and environmentally friendly'.[18]

Monsanto engages in aggressive lobbying when they perceive that their interests are threatened. During the bovine growth hormone controversy, Senator Russ Feinglod of Wisconsin attempted to secure a moratorium on its use until further research was carried out. In advance of a meeting between a Monsanto executive, Tony Coehlo, and the Agriculture Secretary, Mike Espy, in 1993 a memo was written by Dr Virginia Weldon and approved by Monsanto's chief executive, Robert Shapiro, for the purpose of threatening Secretary Espy that, if the Clinton administration did not stand up to persons like Senator Feingold, Monsanto would likely pull out of agricultural biotechnology. The memo then continued with a chilling statement: 'The administration must let socio-economic factors dictate approval of a new product.' Daniel Jeffreys comments: 'In other words, not health considerations, not safety issues, but profits'.[19]

In the light of this kind of behaviour and the corporate culture which supports it, it is not surprising that many people fear that public research institutions and regulatory agencies are working hand-in-glove with transnational corporations (TNCs) to suppress any research that might be construed to be against transgenic crops.

In 1997, *Guardian* reporters John Vidal and Mark Milner also criticized the regulatory agencies in the final article of a four-day special on biotechnology and food. They found:
- a revolving door between the US government and the biotech industry;
- heavy lobbying to rewrite world food safety standards in favour of biotechnology;
- new laws protecting the US food industry from criticism;
- unexpected environmental problems;

---

18. George Monbiot, 'Watch These Beans,' *The Guardian* October 7, 1997.
19. Daniel Jeffreys, art. cit., p. 8.

- local contracts locking farmers into corporate control of production;
- attempts by the world's leading PR firms to massage the debate in favour of genetic engineering;
- the use of world organisations like the World Trade Organisation to challenge governments opposing genetically modified crops;
- consumers being given no effective choice of foods;
- widespread fear that the economies of developing countries will be affected.[20]

The power of transnational corporations to bring pressure on governments and regulatory agencies to take decisions that favour their economic interests was demonstrated once again in February 1999 when the US effectively sabotaged a treaty on biosafety in Columbia. The US refused to allow commodities like soy beans and corn to be included in the treaty and was particularly opposed to transgenic products being labelled. It would seem that the US does not want the consumer to know whether he or she is buying a product that contains transgenic material.[21] In August 2002, Elliot Thomas, of the UK Department for the Environment, Food and Rural Affairs admitted that 'there is enormous international pressure to allow GM crops and seeds in this country ... from the biotech companies. They are going through the national governments and the World Trade Organisation and pressurising the EU'.[22]

THE PRINCIPLE OF SUBSTANTIAL EQUIVALENCE IS TOTALLY INADEQUATE

One reason for the careless US EPA and FDA policy towards transgenic food is that it is based on the inaccurate

---

20. John Vidal and Mark Milner, 'Big Firms Rush for Profits and Power despite Warnings', *The Guardian*, December 15, 1997, p. 4.
21. Jeremy Lennard, art. cit., p. 14.
22. Editorial, 'Sowing Distrust, Europe Goes against the Grain on GM', *The Guardian*, August 17, 2002, p. 19.

premise that genetic engineering is only a minor extension of traditional breeding, not significant enough to warrant a unique policy for this technology.

The biotech industry has opted for the term 'genetic modification organisms' (GMOs) rather than the more accurate term 'genetic engineering organisms' (GEOs). This has been done to try to convince the public that genetic engineering is a simple logical progression from traditional forms of biotechnology as we saw in Chapter Four, pp. 79-80 Throughout this book I use the term 'genetic engineering' except where I am quoting from another source.

The FDA and EPA approach to GE food is based on the notion of 'substantial equivalence'. The concept of 'substantial equivalence' was developed by the Organisation for Economic Co-operation and Development (OECD). It means that, if a new food or food component is found to be substantially equivalent to an existing food or food component, it can be regarded in the same manner from the point of view of safety.

The focus here is not on how the food was produced, whether from natural or genetically engineered seed, but on its chemical nature. Should a chemical analysis of the food or food ingredient find that the product is chemically substantially the same as the naturally produced one, no label should be required on safety grounds.

However, consumer groups and scientists have severely criticised this principle as inadequate. They cite the process that led to the BSE epidemic, and argue that sheep products fed as protein supplement to cattle would probably have passed the substantial equivalence test But the presence of prion proteins led to a public health disaster. Worries about toxicity and allergenicity that have been highlighted earlier might not turn up in laboratory experiments for substantial equivalence which, Dr Mae-Wan Ho points out, are so undiscriminating that 'unintended changes, such as toxins and allergens, could easily escape detection'.[23]

At the international level, the joint Safety Report on genetically-engineered foods issued in 1996 by the Food and Agriculture Organisation (FAO) and the World Health Organisation (WHO) decided that the WHO's *Codex Alimentarius* Commission would decide on the safety of GE foods. Environmental NGOs claim that many of the members of this body are working or have worked for biotech corporations and are therefore biased in favour of the corporations. Representation from consumer and environmental groups is minimal.

Corporate pressures on research institutes or government agencies are not confined to the US as we can see in the case of Dr Arpad Pusztai. Dr Pusztai was born in Hungary and escaped to Britain after the failed revolution of 1956. He received a doctorate in biochemistry at the Lister Institute and in 1963 began to work at the Rowett Research Institute in Aberdeen, Scotland. Working there for the next thirty-five years. he is credited with pioneering work in lectin research. Recognised as a world expert on this little known protein, he has published many books and articles.[24] Dr Pusztai designed the feeding trials for the 1996 Scottish Office-funded GM potato research. He found that ten days after feeding GE potatoes to rats there were major adverse changes in the kidneys, thymus, spleens and guts of the animals. The rats' brain size also decreased. In the light of these disturbing findings, Dr Pusztai called for more research.

With permission from the Rowett Institute Dr Pusztai appeared on the TV programme, *World in Action*, in August 1998 to explain why data gleaned from his own pioneering research on novel methods of biological testing highlighted the need for case-by-case testing of all GE food. The interview went so well that the Director of the Institute, Professor Philip James, phoned Pusztai's wife to compliment him.

23. Mae-Wan Ho, Hartmut Meyer and Joe Cummins, 'The Biotechnology Bubble,' *The Ecologist* May/June 1998: 146.
24. 'The Key Players', *The Guardian*, February 12,1999, p. 7.

Within a few hours the director made a U-turn, from supporting Dr Pusztai to publicly attacking his competence and integrity. Dr Pusztai stresses that he did not claim that all GE foods are unsafe or that biotechnology *per se* is dangerous. He simply stated that his research suggested that GE foods may pose dangers to human health, and that more research is needed. He was concerned that GE foods already in the food chain had not been subjected to the kind of test which he used in his research.

What followed seems bizarre. Dr Pusztai was suspended, his phone calls were redirected to the director's office, and his emails were intercepted. Professor James threatened Pusztai with legal action if he spoke about his work to anyone outside the institute. Remarkably, when the Rowett Institute did an audit on Pusztai's work, none of the nutritionists at the Institute was appointed to the audit committee. Most amazing of all, Pusztai was not given an opportunity to explain his work to the committee and challenge his detractors.

Dr Pusztai believes that he was sacked at the behest of the biotech industry. These corporations have extraordinary power in the US and the UK. It would seem that they saw his work as a threat to their profits and therefore he was expendable. If such punishment could be meted out to someone like Dr Pusztai, who had an unblemished record of scientific achievements, one can only imagine what would happen to a young scientist whose research raised questions about the safety of GE foods.[25] As the controversy raged about Pusztai's sacking it emerged later that the Rowett Institute had received a £140,000 grant from Monsanto.[26]

Twenty-two well-known scientists wrote a letter of support for Dr Pusztai's work and called for his reinstatement. Dr Vyvyan Howard, head of research and toxicology at Liverpool

25. Arpad Pusztai. 'Academic Freedom: Is It Dying Out', *The Ecologist*, April 2000, pp. 26- 29.
26. Christopher Leake, 'Gene Lab Took Food Giant's Cash Gift', *The Mail*, February 14, 1999, pp. 1 and 5.

University, stated that he found Dr Pusztai's data sound and was sure that it would be well reviewed by his peers and should be published. He told the *Express*: 'We are at a loss to explain why the Rowett Institute came to the conclusions it did'.[27] Dr Stanley Ewen of Aberdeen University repeated Pusztai's experiments and came to a similar conclusion.[28]

What happened to Dr Pusztai raises serious questions about the independence of science now that so much research is under the control of corporate paymasters. Writing in a book called *Safe Food*, Dr Pusztai's review of the literature led him to conclude that some GE foods may be unsafe. The biotech companies always claimed that DNA would be broken down in the gut soon after GE food was consumed. Now research is indicating that genetically engineered DNA in plants can be taken up by gut bacteria in both animals and humans. This raises the fear that foreign genes which have been inserted into crops to confer on them properties like antibiotic resistance, could pass this ability on to bacteria, making them also resistant to antibiotics. If there is even the slightest possibility that GE food is unsafe, then the precautionary principle ought to dictate that it not be introduced into the food-chain.[29]

The fact is that GE food is new and has the potential to do a lot of damage. Therefore, each GE product ought to be treated as the novel process that it is. Regulations ought to be much more demanding and rigorous. Independent verification must be built into the process if it is to win public trust. Adequate resources need to be put into the regulatory agencies so that the research is thorough. The FDA must require biotech corporations to notify them when any genetically

---

27. John Ingham and Lorna Duckworth, 'This Scientist Revealed the Perils of GM Food. Now He Has Been Gagged for Life', *The Express*, February 13, 1999, p. 7.
28. Ibid.
29. 'Scientist Who Pressed GM Panic Button Raises New Food Health Fears', *The Sunday Times*, May 4, 2003, p. 12.

engineered food goes on the market. In the event of problems, this would provide a trail for scientists, medical personnel and regulators to follow in order to determine the origins of an unsafe product.

REGISTRATION, SEGREGATION AND LABELLING

Opinion polls in Europe and elsewhere indicate that the public want clear and informative labelling of genetically engineered foods. In 1997, polls in Germany showed that 90 percent of those questioned wanted more information and clear labelling.[30] Even in the US, in the early 1990s the vast majority of people thought that labelling was very important. Similarly in Australia and Canada: 90 percent of consumers polled want specific labelling to allow shoppers to make an informed choice.[31] It is clear from a *Guardian* poll of June 1998 that people want to know whether the food they eat is derived from genetically engineered products: in response to the question whether 'foods that have been genetically modified should be clearly labelled', 96 percent answered Yes. In 1997, a similar opinion poll in the US, commissioned by Novartis, found that 97 percent wanted all biotech food labelled. The Irish Department of the Environment's Consultation Paper recognises that 'from a consumer point of view, the information currently being provided is inadequate to facilitate clear choices on whether or not to purchase products containing GMOs or products using genetic modification techniques'.[32]

A basic step in honouring this consumer preference is to segregate genetically engineered foods from natural foods at source. Many of the corporations that produce genetically engineered foods are opposed to segregation and labelling. They claim that it would be too expensive since it would

30. Roger Crowe and John Vidal, 'Backlash against Genetic Farming', *The Guardian*, December 17, p. 4.
31. Ibid.
32. *Genetically Modified Organisms and the Environment: A Consultation Paper*, Department of the Environment, Dublin, August 1998, p. ix.

involve using different containers, trucks and warehouses. They protest that their products are safe. Without segregation, for example, tomato paste from genetically engineered tomatoes would have to be labelled, but a processed food like lasagne that contained genetically engineered tomatoes would need no label. Labelling of all genetically engineered food products should be required by law and not left to the retail outlets. A report from an Irish government panel, tasked with evaluating the pros and cons of GE food, supported the biotech opposition to segregation despite the protest of consumer groups that they have a right to know where and how food is produced. The three-person panel, composed of Turlough O'Donnell QC, Professor Dervilla Donnelly (UCD) and Seán Cromien (former Secretary-General of the Department of Finance), wrote that the demand by environmentalists and others for clear labelling

> ignores global trade considerations as well as the economic realities of mass producing agricultural and agrifood products at reasonable prices. The policy of mixing conventional and genetically-modified produce in non-EU nations, particularly countries with whom Ireland and the EU have important trade relations, such as the United States and Canada, is a reality that must be faced and which requires a more practical policy approach than segregation.[33]

The subtext is very clear: money talks; economic considerations take precedence over human health.

Segregation and clear labelling are the minimum requirements necessary to ensure product safety and to protect a consumer's right to choose whether or not to purchase these products. Consumer groups all over Europe and the US have called for such a system. Once again, in the matter of labelling the public in the US are poorly served by the regulatory

33. 'Genetically Modified American Pie', *The Phoenix*, October 22, 1999, p. 5.

agencies. In Chapter Four I described how people allergic to nuts were found to be allergic to a transgenic soybean that contained a gene from a Brazil nut. Reflecting on this discovery, the editor of the prestigious medical magazine, *The New England Journal of Medicine*, chided the FDA in the US for its unwillingness to demand verifiable and across-the-board labelling. For the editor, it appeared that the FDA 'favours industry over consumer protection'.[34]

The decision by some retail outlets to label with the words 'may contain genetically modified products' means little. If there is no segregation almost all processed food might contain some genetically modified substances. In practice, this would prevent the consumer from choosing foods that are not genetically engineered.

THE ECONOMIC AND POLITICAL POWER OF MONSANTO

The stakes are high for a company like Monsanto. It was originally a chemical company, founded in St Louis, Missouri, in 1901. By the 1940s, it had become an important player in the plastics and synthetic fabrics industries. In the 1960s and 1970s, it produced the lethal herbicide Agent Orange which the US military used as a defoliant in Vietnam. The effects of this toxic chemical are still being experienced by US war veterans, Vietnamese soldiers and civilians. After the Vietnam War, Monsanto paid $80 million in an out-of-court settlement with US veterans. The Vietnamese have yet to be properly compensated.

The company was also a major producer of polychlorinated biphenyls (PCBs) which are used as lubricants and coolants in various industries. Many residents of Anniston, Alabama, feel that they have been poisoned by PCBs that Monsanto manufactured there. Three thousand five hundred of them took legal action against Monsanto. What was at issue was whether

---

34. Jeremy Rifkin, *The Biotech Century*, 1998, London, Victor Gollancz, p. 105.

Monsanto knew about the harmful effects of PCB contamination and whether they could have protected the community.[35] At the trial which started in August 2001, lawyers for the citizens claimed that Monsanto routinely discharged toxic waste into a west Anniston creek and dumped millions of pounds of PCBs into oozing open-pit land-fills. Thousands of pages of Monsanto documents – many emblazoned with warnings such as 'CONFIDENTIAL: Read and Destroy' – show that for decades, the corporate giant concealed the danger.[36]

In the early 1990s the chief executive, Bob Shapiro, came to believe that the future for Monsanto lay with biotech products rather than chemicals. So, it began to reinvent itself as a biotech company. Worth an estimated $400 billion a year, the biotech market was expected to rise dramatically.[37] Many commentators believed that Monsanto was about to become the Microsoft of the biotech world.[38] After buying out or taking control of many small, innovative biotech companies, including Delta Pine, Monsanto turned its attention to large seed distribution corporations. In 1997, it bought Holden's Foundation Seeds for $1.2 billion. A year later, in June 1998, it paid a record £843 million for Cargill Incorporated. This huge agribusiness has sales and distribution networks in fifty-one countries on four continents. This acquisition gave Monsanto, a chemical company, huge control of global seed markets. Shapiro saw all these companies as conduits for distributing their GE seeds. The stock market responded by pushing Monsanto's share from $11.50 to $45. It was given a group market capitalisation of $26.7 billion. Some months

---

35. Nancy Beiles, 'People v Monsanto', *The Guardian* (Supplement), June 5, 2000, pp. 2 and 3.
36. Michael Grunwald, 'Monsanto Hid Decades of Pollution: PCBs Drenched Ala. Town, But No One Was Ever Told', *Washington Post*, January 1, 2002, p. A01.
37. John Vidal and Mark Milner, 'Big Firms Rush for Profits and Power despite Warnings', *The Guardian*, December 15, 1997.
38. George Monbiot, John Farvey, Mark Milner, John Vidal, 'Biotech Firm Has Eyes on All You Can Eat', *The Guardian*, December 5, 1997, p. 3.

later, Monsanto shares jumped to $62.

Commentators like George Monbiot of the *Guardian* watched in horror at what was happening. He was astonished by the speed at which 'a tiny handful of companies is coming to govern the development, production, processing and marketing of our most fundamental commodity: food. The power and strategic control they are amassing will make the oil industry look like a corner shop'.[39] He was frightened by the possibility that within a few years the world's food supply could be dominated by eleven or fewer giant Northern controlled agribusiness corporations.

Monsanto was wielding more and more political power in the US, with friends in both the Democratic and Republican parties. The company has made huge donations to both, and has paid lobbyists to represent its interests at both state and federal level. It has even made financial contributions to members of Congress who sit on food safety and regulatory committees. Within the US political system, this, unfortunately, is quite legal. Mickey Cantor, who was the chief US negotiator during the Uruguay Round of the GATT negotiations and the chairman of Bill Clinton's 1992 presidential election campaign, now sits on the board of Monsanto.[40] According to Betty Martini, of the consumer group Mission Possible which monitors Monsanto's activities in the US, 'the Food and Drug Administration, which regulates the US food industry, is so closely linked to the biotech industry that it could be described as their Washington branch office'.[41] John Vidal also observed that there is a 'constant exchange of staff between the government, the company and the regulatory bodies'.[42]

---

39. George Monbiot, *The Guardian*, September 17, 1997.
40. Julian Borger, 'Why Americans Are Happy', *The Guardian*, February 20, 1999, p. 4.
41. John Vidal, 'Biotech Food Giant Wields power in Washington', *The Guardian*, February 18, p. 8.
42. Ibid.

Monsanto's influence is felt beyond the boundaries of the US. In January 1999, *The Sunday Tribune* reported that during the visit of the Taoiseach, Bertie Ahern, to the US in March, 1998, leading figures in the US administration, including Sandy Berger, director of the US National Security Council, used the Taoiseach's visit to try to influence Ireland's vote on the upcoming decision about planting crops engineered for insect resistance in the EU. Politicians, including Senator Christopher Bond, collared Ireland's Prime Minister on the subject, according to a report in the *St Louis Post-Dispatch*. Some commentators described as worrying and frustrating 'the access Monsanto had to the Taoiseach during the visit'.[43] It would be very difficult for Mr Ahern to brush aside overtures from US administration personnel or politicians, given the pivotal role the US played in the Northern Ireland peace process.

Bertie Ahern directed the then Minister for the Environment, Noel Dempsey, to instruct an Irish official attending a crucial EU meeting to support an application to import genetically modified corn. Minister Dempsey told the Dáil: 'my overall preference was to abstain in individual votes pending the completion of the national consultation process'.[44]

Did hard-nosed lobbying by the US and Irish biotech industries account for the fact that Fianna Fáil (the largest party in Ireland) quietly dropped its hostility to genetic engineering once it entered government in 1997? While in opposition, the spokespersons for the environment, Noel Dempsey, and for agriculture, Joe Walsh, published a policy paper in which they stated that 'it would be premature to release genetically modified organisms into the environment or to market food which contains any genetically modified ingredient.' The reason given for this stance was:

43. Claire Grady, 'Ahern Lobbied on Modified Crops', *The Sunday Tribune*, January 3, 1999, p. 3.
44. 'Genetically Modified American Pie', *The Phoenix*, loc. cit.

Current scientific knowledge is inadequate to protect the consumer and the environment from the unpredictable and potentially disastrous side effects which may appear immediately or at any time in the future. The principle of substantial equivalence governing the testing and labelling of genetically modified food requires that only known potential hazards are tested. Therefore unexpected toxins and allergens will only be identified when a unique health crisis occurs. This will be too late for the victims. Fianna Fail will not support what amounts to the largest nutritional experiment in human history with the consumer as guinea pig.[45]

Once in government, this promise was quickly forgotten, under pressure no doubt from biotech companies.

The rapid spread of GE soybeans, cotton and corn is a testimony to the economic and political pressure which the Clinton and Bush administrations have brought on other countries. In 1996, only 1.7 million hectares had been planted with GE soybeans. By 2001, that had jumped to 52.6 million hectares. In April 2002, Brazil was the only major soybean producing country that continued to ban GE varieties. The decision has benefited Brazil economically because many consumers prefer GE-free soya. Brazil's share of the global soybean market jumped from 24 to 30 percent while the US share has fallen from 57 to 46 percent over the same period.

This development has not pleased either the US government or Monsanto. In January 2002, Anthony Harrington, a former US ambassador to Brazil, then working as a consultant for Monsanto, had a meeting with President Fernando Cardoso to try and promote GE soya. The strategy was simple. If Brazil were to grow GE soybeans, consumers in Europe who were demanding non-GE soybeans would not be able to get them. Such bullying tactics by the corporations are designed to

---

45. Taken from a press release entitled 'Fianna Fáil Position on Genetic Engineering of Food', April 26, 1997.

ensure that the consumers do not have the right to choose non-GE food.[46] It is particularly offensive when a government like the US which proclaims individual freedom of choice colludes with corporations to deny people that freedom.

Rejection of GE maize by the EU and other countries has also had a serious impact on US maize exports. While only 35 percent of the maize grown in the US is genetically engineered, because only 2 percent is segregated, 98 percent of US maize is now banned from the EU and other markets.

By the middle of 1999 it was clear that there were more problems with GE food than the biotech industry and, especially Monsanto, had anticipated. Europeans, especially British consumers and French farmers, were not favouring GE food; and the world was not queuing up at super-markets demanding GE foods. Markets for GE corn and soya began to dry up in Japan, Taiwan and Mexico. In fact, wherever people had a choice, they were shunning GE food. Monsanto realised that they had struck the wrong chord in their advertisements in Europe. They needed a fresh start to win over skeptical Europeans. One element in this new ploy saw Monsanto's chairman Bob Shapiro dialoguing with a Greenpeace audience in a conference call in November 1999 when he confessed: 'Our confidence in this technology [GE] and our enthusiasm for it has, I think, been widely seen – and understandably so – as condescension or indeed arrogance'.[47]

The financial markets began to take note of such developments after the EU moratorium on planting GE crops, agreed at meeting in Brussels in June 1998. As a consequence, the bubble burst on shares in biotech companies. Europe's biggest bank advised investors to sell shares in biotech companies because consumers were not buying GE products.[48] Monsanto

46. Sue Branford, 'Sow Resistant', *The Guardian*, (Supplement) April 17, 2002, pp. 8 and 9.
47. Julian Bolger, 'How the Mighty Have Fallen', *The Guardian*, (Supplement), November 22, 1998, p. 2.
48. Paul Brown, 'GM Investors Told to Sell Shares', *The Guardian*, August 25, 1999, p. 1.

took the hardest hit because it had invested so heavily in GE technology and distribution outlets. It was clear that the gamble which Shapiro took in the early 1990s to put all his eggs into the biotech basket, was not going to succeed. The company began to experience cash-flow problems as expected revenue was not being achieved. Its share value fell to $38 dollars and there was speculation whether the company would be forced to sell its pharmaceutical subsidiary, GD Searle & Co.[49] As share value continued to drop it was forced to merge with Pharmacia and Upjohn. In December 1999, Bob Shapiro was ousted from the company.[50]

It is much too early to write off Monsanto or any other biotech company. In the US, such companies still wield enormous power. Operating in over 60 countries, Monsanto is still pursuing an aggressive pro-GE food policy globally as we will see when I discuss the commercial planting of Bt corn in the Philippines (Chaper Six).

As we saw in Chapter One, Monsanto is well represented in President Bush's cabinet. The Secretaries of Defence, Health and Agriculture, the Attorney-General and the chairperson of the House Agriculture Committee have had connections with Monsanto or the wider biotech industry. Writing in February 2002, Charles Lewis, director of the Centre for Public Integrity, said: 'It looks like Monsanto and the biotechnology industry has the potential to bring undue influence on the new government'.[51] Linda Fisher, a former executive at Monsanto, became deputy administrator of the Environmental Protection Agency.

In 2003, Monsanto wants the US government to influence the World Trade Organisation to put pressure on Europe to open its markets to GE products from the US. The European Commission has rejected the US decision to go to the WTO

49. Jane Martinson, 'Monsanto Pays GM Price', *The Guardian*, December 21, 1999, p. 19. Julian Bolger, art. cit., p. 2.
50. Jane Martinson, Ibid.
51. John Vidal, 'George Bush's America', *The Guardian*, February 1, 2002.

as misguided and unnecessary. While one would like to see a complete moratorium on the release of GE plants or food, at least the EU regulatory process is much more rigorous than that in the US. The rules governing the release of GE organisms into the environment are set out in Directive 2001/18/EC.

If a company wishes to produce and sell a GE product in the EU, an application must be made through a Member State. The application must contain a full environmental risk assessment. If the national authorising body (in Ireland the Environmental ProtectionAgency, EPA) rules that the product can be safely allowed into the marketplace, this Member State informs other Member States through the EU Commission. If there are no objections, the body that carried out the original assessment gives permission for this product to be placed on the market. If objections are raised, the decision has to be taken at EU level. The EU Commission will normally seek the opinions of its Scientific Committees, composed of independent experts who are qualified in a variety of disciplines involving food, including medicine, nutrition, toxicology, biology and chemistry. If the scientific opinion is favourable, the Commission proposes a draft decision to the Regulatory Committee, which is composed of representatives from the Member States. If the Regulatory Committee's decision is favourable, the Commission allows the product to be sold right across the EU. If the Regulatory Committee dissents, the decision is submitted to the Council of Ministers where it can be adopted or rejected by qualified majority. If the Council does not act within three months the Commission may decide.

The public can gain access to the various positions on a website: www.gmoinfo.jrc.it/ As of May 2003 eighteen GE products have been authorised for sale in the EU. Regrettably, more are expected in the next few months.

Every product that contains more than 1 percent of genetically modified material must be labelled, be it manufactured within the EU or imported from a non-EU country. EU rules regarding labelling are set out in EC 258/97 on Novel Food Ingredients. Once again, the first step takes place in a Member State. The competent body in that state (the EPA in Ireland) assesses the application to market the GE food. Once again, if the decision is favourable the State informs other Member States through the EU Commission. If there are objections, a decision will have to be taken at the Community level. The Commission consults the Scientific Committees on the question of public health. If the Scientific Committees consider that there is no risk to public health, the food can be sold across the EU.

Directive 2001/18/EC which obliges Member States to take all the necessary measures to label GE organisms in food or any other products, came into effect on October 17, 2002.

During the framing of these regulations, governments, the competent authorities in each state (EPAs) and the biotech industry were in constant communication. The corporations seemed to have scored a significant victory when they won a derogation from the full Novel Food Regulations for foods derived from GE organisms but which no longer contained GEOs. Here they employed the 'substantially equivalent' argument that if the food was similar to existing food in respect to composition, nutritional value, metabolism, intended use and the level of undesirable substances, the company would only have to notify the Commission that they were placing the product on the market. With this notification they would have to furnish scientific justification for the claim that the food is 'substantially equivalent'. But as we saw earlier, the arguments behind the notion of substantial equivalence are very threadbare.[52]

Traceability is very important if we are to avoid polluting

52. 'European Commission Regrets the US Decision to File WTO Case on GMOs as Misguided and Unnecessary', www.europa.eu.int/rapid/start/cgi/guesten.ksh p. 2.

the human food chain. In 2000, GE corn not approved for human consumption was found to have entered into about 300 food products, including the snack bar Starlink. More than 300 branded products had to be withdrawn from supermarkets by the US authorities.

Finally, the movement of GE food or organisms internationally is governed by a number of agreements. Included in these are the WTO Sanitary and Phytosanitary Measure (SPS) Agreements, the WTO Technical Barriers to Trade (TBT) Agreement, the Cartagena Protocol on Biosafety, and the Convention on Biological Diversity. The Cartagena Protocol regulates the movement of GE organisms or food across country boundaries. It has been signed by 102 countries, including the EU Member States. Unfortunately, the US has not signed up to this protocol or the Convention on Biodiversity. The Cartagena Protocol will come into effect when 50 countries have ratified it. Forty-two countries had ratified it by the end of January 2003.

The EU has been critical of the fact that the US was sending GE corn to countries in Africa which were facing starvation. Between September 2002 and March 2003, the US committed 499,000 tonnes of GE maize to serve as food aid. Many saw this as an attempt by the US authorities to force GE food into Africa.[53] Zambia refused to accept such food aid. The US sponsored a study trip for Zambian scientists to the US and Europe so that they could see the benefits of GE maize and communicate that to their government. The Zambian scientists met with European colleagues under the auspices of the EU Commission. On returning home they advised against introducing GE maize in Southern Africa.[54] The US administration blamed the Europeans for influencing the African scientists. The EU dismissed the claim, saying that, as far as they were concerned, there was no inherent danger in GE

---

53. Ibid p. 6.
54. Ibid p. 8.

foods, but that sovereign governments had the right to legislate and regulate this important area.

It is becoming increasingly clear that the US intention is to force GE food into every country in the world. According to Don Westfall, vice-president of a biotech consultancy firm, Promar International, the US plan 'is that over time the market is so flooded [with GM] that there's nothing you can do about it. You just sort of surrender'.[55]

In this chapter I argued that a further reason for being cautious about GE food is the incompetent, poorly resourced and corporate-friendly regulatory climate – especially in the US. I also returned to a theme which runs right through this book, namely the power of large corporations to influence governments to do their bidding, even when it is not in the best interest of ordinary citizens. Monsanto and others have put pressure on governments around the world to open their markets to GE food. They are also currently attempting to use the WTO for the same purpose. But the corporate reach moves far beyond the commercial and political sphere – as Dr Arpad Pusztai learned when some of his research raised serious questions about GM potatoes, he was removed from his position and his scientific competence attacked.

The more positive feature of this chapter is that consumers have demanded more protection from their governments, especially in Europe. Now, food will have to be labelled in a credible way. This, I believe, will put the brakes on the global reach of GE foods because, given a real choice, the consumer will reject it.

55. Editorial, *The Ecologist*, July/August 2002, p. 4.

SIX

Ethics and Genetic Engineering

GENETIC ENGINEERING is a new and powerful technology. It gives enormous power to commercial organisations like transnational corporations (TNCs) to transform, not just human life, but life itself.

In discussing the ethics of genetic engineering it is essential to develop an appropriate ethical framework for this new and powerful technology. This will demand a major shift away from the almost exclusively human or homocentric focus that has been so pervasive in the Western ethics and the wider cultural traditions for almost two thousand years.

WESTERN ETHICAL AND CHRISTIAN TRADITION ARE HUMAN-CENTRED

Aristotle, whose impact on Western thought is enormous, held that since 'nature makes nothing without some end in view, nothing to no purpose, it must be that nature has made [animals and plants] for the sake of man'.[1] This idea, that animals and plants are created for humankind – either by God or the processes of nature – has dominated Western attitudes to animals, plants and the rest of creation for many centuries. Thomas Aquinas shared this view that the rest of creation was there to benefit humankind.[2]

From this viewpoint, since animals and plants exist for human beings, our behaviour towards them is not governed

---

1. Aristotle, *Politics*, Harmondsworth, Penguin, 1985 edition, p. 79.
2. Thomas Aquinas, *Summa Contra Gentiles*, III, Chapter CXII, reprinted in *Animal Rights and Human Obligations*, 2nd edition, T. Regan and P. Singer (eds), Prentice Hall, New York, 1989, p. 56. Cited in Celia Deane Drummond, *Theology and Biotechnology*, Geoffrey Chapman, London, 1997, p. 98.

by moral considerations. It is only in the past decade that the cruelty involved in factory farming or blood sports has been discussed from an ethical perspective. Even then, the proscription on cruelty towards animals arises, not so much from inherent rights that animals might have, but from the understanding that any form of cruelty is unbecoming and, therefore, unethical for rational beings.

It is also true that certain elements within the Judeo-Christian tradition have strongly reinforced the Aristotelian legacy. This is true when one considers the traditional interpretation given to the biblical text, 'Increase and multiply and dominate the earth' (Genesis 1:26-28). The text is often interpreted, mistakenly according to contemporary scripture scholars, as giving humans a licence to dominate the earth and to do whatever they wish with animals and plants.

The historian, Keith Thomas, points out that at the beginning of the sixteenth century, just as modern science was finding its feet, neither Western literature nor the theological tradition ascribed any intrinsic meaning to the natural world or accorded it any rights apart from its role in serving humankind.[3] From the theological perspective, it was argued that humans had intrinsic value because they were made in the 'image and likeness of God' (Genesis 1:26). Their role was to be 'masters of the fish of the sea, the birds of heaven and all living animals on the earth' (Genesis 1:28). No other creature bore this image of God (*imago Dei*) stamp. Animals and plants were viewed as lacking rational faculties, self-consciousness and often even sentience, and hence as having no intrinsic worth in themselves. They only had instrumental value; their role was to serve the needs of humankind for the necessities of life and they could also be used for entertainment.

Even though the Catholic Church has attempted to develop a creation theology in recent years and, in some pro-

---

3. Keith Thomas, *Man and the Natural World*, New York, Pantheon Books, 1983, p. 35.

nouncements has begun to accept that other species have intrinsic value, many Church people still work out of an anthropocentric perspective. Speaking at a conference on biotechnology in October 2002, Bishop Elio Sgreccia, vice-president of the Pontifical Academy for Life, said: 'There are no impediments to animal and vegetable biotechnologies. The latter can be justified with the motive that they are for the good of man. God has conceived animals and vegetables as good creatures for man's needs.' The bishop added an unspecified caution when he stated that 'God has also given man the task and responsibility to govern creation, which implies a grave responsibility ... therefore, the use of plants and animals is legitimate, but it does not represent an absolute right.' According to the bishop, the Church has an open but conditioned position: 'For this reason, we ask for sales to be accompanied by a label [mentioning GMOs] and their total availability for developing countries, in keeping with criteria of solidarity and justice'.[4]

The bishop focused completely on the impact of this new technology on humankind, even when he championed labelling and equity for hungry people in the Third World. The rights of other species not to be subjected to cruel experimentation and to have their specific genetic integrity respected by human beings are not even raised.

It is true that within the Judeo-Christian tradition there is a strand that sees humans as stewards of creation (Genesis 2:15). However, as Clive Ponting points out in his *Green History of the World*, 'although the idea that humans have a responsibility to preserve the natural world of which they are merely guardians can be traced through a succession of thinkers, it has remained a minority tradition'.[5] Unfortunately, St Francis's kinship with brother Sun, Sister Moon and all creation was very much a minority position. His fraternal attitude did not

   4. www. zenit.org Rome, October 9, 2002, p. 2.
   5. Clive Ponting, *A Green History of the World*, London, Sinclair, Stevenson, 1991, p. 142.

inform the Western approach to nature. In fact, it did not even survive in any effective way within the Order which he founded.

THE SCIENTIFIC REVOLUTION WIDENED THE CHASM BETWEEN HUMANITY AND THE REST OF CREATION

The gulf between humans and the rest of creation was widened further by the insights of many of the people who shaped our modern scientific, economic and social world. In this very formative period in human-earth relations, when the foundations of the modern scientific and industrial society were being shaped in the works of people like Francis Bacon (1561-1626), René Descartes (1596-1650) and Isaac Newton (1642-1727), all rights were ascribed to humans. In the words of Descartes, the goal of human knowledge and technology was so that humans might become 'the masters and possessors of nature'.[6] Furthermore, philosophers like Thomas Hobbes, John Locke and Jeremy Bentham dismissed the medieval view of the cosmos as organic and substituted instead a mechanistic view of nature and its law. For these men, the best way to understand the Cosmos was to see it as a giant clock. Newton believed that the laws of motion which he discovered, proved that the same universal laws that governed the smallest clock also governed the movements of the earth, the sun and the planets.

For Descartes, animals were *res extensa*, little more than mechanised entities without any interior quality or soul. Only *res cogitans* or humans, as conscious beings, were endowed with souls and, therefore, could be considered to have moral value. Animals had no such intrinsic value and could be treated in any way that might serve human ends, no matter how cruel and degrading that might be. Moreover, these men of the Enlightenment viewed science and its handmaiden,

---

6. Stephen Mason, *A History of the Sciences*, Collier Books, New York, 1962, p. 27.

technology, as a tool designed to give humans the power to dominate and manipulate the earth in whatever way they saw fit, in order to promote human well-being and betterment.

Genetic engineering fits comfortably into this mechanistic world-view. Scientists, working in the field of genetics and biotechnology, discovered the insights and technology which people like Francis Bacon dreamed about in his book, *New Atlantis* (1627), published after he died. This new technology gives humans the capability to manipulate the building blocks of life in order to reshape the natural world in a most extraordinary way. Most of the questions surrounding genetic engineering deal with whether it will be damaging to human health or the environment. In general, the wider ethical questions are often simply ignored.

A HIERARCHY OF MORAL QUESTIONS
SURROUNDING GENETIC ENGINEERING

The fundamental question is: Does genetic engineering respect the intrinsic rights of other creatures? Having explored that question, one might then move on to some of the other ethical questions associated with the technology. These would include the question whether the risks to the environment and human health from genetic engineering are serious enough to warrant a moratorium on deliberate release of genetically modified organisms at this time. Within the context of the human community, will genetic engineering further widen the gap between rich and poor in our contemporary world? Will genetic engineering respect the rights of Third World people who have promoted biodiversity over thousands of years? Or will it facilitate the plunder of these countries which are much richer biologically than many Northern corporations? This phenomenon is called biopiracy. Finally, the push to patent genetically engineered organisms raises the basic ethical question: Is it proper to patent or claim ownership over living organisms? I will deal with the patenting issue in the next chapter.

These questions are not an argument for stopping research on genetic engineering in laboratories. They mean that those who wish to engage in deliberate releases will have to demonstrate that the genetic rights of the organisms they changed have not been violated, that their products will not damage human health or the environment, and, finally, that transgenetic seeds or animals will benefit the poor of the world and not just increase the profits of biotech companies. One reason for extreme caution is that if something does go wrong it will be impossible to recall the organisms that are multiplying in the environment. Genetic engineering deals with organisms that produce, mutate and interact with other organisms in the environment. I will return to this point later.

AN ECO-CENTRED MORAL FRAMEWORK

Before looking at suitable moral frameworks for assessing the morality of genetic engineering it might be worthwhile to consider what has happened to the North Atlantic salmon. A report published in 1997, commissioned by the Marine Institute of Ireland, discusses *The Nature and Current Status of Transgenetic Atlantic Salmon*. As a result of introducing growth hormone genes into a wild North Atlantic salmon the transgenic fish grows rapidly and reaches enormous size. Studies show that within a period of 14 months the transgenic salmon was 37 times heavier than the average.[7] These increases will probably make enormous profits for the company producing the salmon. The cost to the salmon is horrendous. In its technical and unemotive language the report notes that the experiment produces 'profound morphological abnormalities in the progeny of the transgenetic salmon. These included a disproportionate growth of the head and operculum cartilage, disimproving appearance and leading ultimately to respiratory problems'.[8] In ordinary language, what happened

7. T.F. Cross and P.T. Galvin, *The Nature and Current Status of Transgenetic Salmon*, Dublin, Marine Institute of Ireland, 1996, p. 6.
8. Ibid.

was that the physiognomy of the fish was so distorted by the genetic manipulation that it died a painful death because it could not breathe. The report recommended that permission to grow GE salmon should not be given in Ireland because of the genetic consequence of the GE salmon breaking loose and contaminating the ordinary north Atlantic salmon stock. The report never raised the basic ethical question: Do humans have the right to interfere in such an intrusive way with the genetic integrity of this species of fish?

Another example is the research and experiments on pigs at Newcastle University in Britain. Ruminant animals, like cattle, produce enzymes in their gut that break down cellulose plants into basic sugar components which are then assimilated by the animal. Now scientists are experimenting with introducing cellulose genes directly into non-ruminant animals, like pigs and chickens. Their aim is to produce the 'grazing' pig. Whether this would be good for the pigs, given that the rest of their physiology does not suit grazing behaviour, or for the soil structure, is not at all clear.

In recent years, experiments have also been carried out on many other animals in an effort to improve livestock or develop cheap ways of producing drugs. Animal rights groups, like Compassion in World Farming, are rightly concerned about the suffering which genetic engineering techniques inflict on animals. As we saw in Chapter Three, current genetic engineering technology is imprecise because the expression of the gene depends on the promoters, enhancers and silencer genes. As a result, all kinds of abnormalities have occurred, including loss of limbs, and brain defects. In many situations the transgenic animal does not pass on the desired gene to its offspring; so, repeated experiments are necessary in order to develop the desired line for breeding purposes. Jeremy Rifkin is very conscious of the fact that GE technology, as it stands now, has the potential for both increasing animal suffering and creating abnormal creatures. He writes: 'The larger lesson is that the complex and multiple interactions between the inserted gene and the chemical activity of the host animal are, for the most part, unknowable and unpredictable and can result in all sorts of novel and even bizarre pathologies in the creature'.[9]

An ethical assessment of genetic engineering or cloning needs to ask the following question: Should we, as one species among millions of others, be engaging in such intrusive experiments changing the genetic integrity of other species? What human need would justify such intervention? Would the desire to produce an animal with profitable economic traits like increased growth performance, leaner meat and greater weight justify the operation?

I believe that our anthropocentric Western scientific values and ethical norms are not capable of addressing these vital contemporary moral issues in any comprehensive or effective way. Our moral compass is puny even while our technologies have the potential to do damage on a planetary scale. Even the 'minimum ethical consensus' proposed by the theologian Hans Küng appears to be mainly human-centred. It includes the fundamental right to life (human), just treatment from the State, and physical and mental integrity. By examining various religious traditions and seeing what ethical values they share, Küng is attempting to develop a global ethic which would allow humans to live and work together, based on the lowest common denominator of shared ethical values. While we yearn for a more peaceful world in which people can respect cultural and religious differences, starting on a lowest common denominator basis does not seem to be a very viable or creative way to go about developing a global ethic.[10]

---

9. Jeremy Rifkin, *The Biotech Century*, 1998, London, Victor Gallanz, p. 97. Rifkin cites Gill Langley, 'A Critical View of the Use of Genetically Engineered Animals in the Laboratory, in Wheale and McNally (eds.), *Animal Genetic Engineering*, London, Pluto Press, pp 184-8.
10. Hans Küng, *A Global Ethic for Global Politics and Economics*, 1997, London, SCM Press, pp. 91-113.

The relationship between humans and other creatures is not on Küng's ethical landscape at all. Yet in 2003 this relationship is a matter of life and death for many species, possibly, even the human race. Küng provides a hint at how we might attempt to construct an adequate ethical framework for contemporary problems like genetic engineering. He suggests that a global ethic be 'related to reality'. While accepting this position, I think it should be broadened in a way that situates the human story and reality within the larger story of the earth and the universe.

In this view, a satisfactory ethical framework must be based on our contemporary understanding of the relationship between humans and the rest of the natural world, not on the mechanistic outlook of Newton or Descartes. In the scientific world of the 1980s and 1990s, the mechanistic view has being challenged by physicists and biologists. John Polkinghorne, a theologian and former professor of theoretical physics at Queen's College, Cambridge, insists that 'the world is no mere mechanism. It has a flexibility, a suppleness, within its process, a freedom for the whole universe to be itself, a freedom for us to act within that universe of which we are a part'.[11] Our evolutionary history makes it very clear that humans are not disconnected from the rest of nature. Rather, we are an integral part of the community of living beings and non-living reality.

Humankind evolved with other creatures during the past two million years and we are dependent on plants and animals for our survival. We are part of the web of life. The well-being of the human species depends on the well-being of the whole fabric of nature. If we damage that in an irreversible way, we damage ourselves. So, even from the perspective of enlightened self-interest, we ought to respect the community of living beings as well as the air, water and soils of the earth to

11. John Polkinghorne, 'The Unity of Truth in Science and Theology', in *Science and the Theology of Creation*, Church and Society Documents, No. 4, August 1988, p. 31.

secure our own future.

Much of the moral debate in this area concentrates on the impact of genetic engineering on human beings.[12] It focuses on whether it will benefit or damage human health. Genetic engineering techniques make it possible to alter in a significant way the genetic integrity of any species, be it a bacterium, plant or animal. But is it ethically right to do this particularly if the modification is harmful to the animal, as in the case of the transgenetic salmon?

Viewed through an exclusively anthropocentric moral framework the answer to this may well be: Yes. Charles McCarthy, an ethicist with the Kennedy Institute for Bioethics at Georgetown University in Washington D.C., writes: 'in a utilitarian context, efficiency in food production and ability to compete for world markets stand as high values which must be weighed against our recognised obligations to provide for the interests of the animals'.[13] In general, the human-centred argument usually wins out.

As I have stated earlier, the moral framework which we inherited from the Greek and Roman culture and a segment of the Judeo-Christian tradition, seems no longer adequate in itself for assessing the ethics of a complex issue like genetically engineering other creatures. The attempt to widen the moral universe beyond the human domain to include the rights of other species has been under way on the margins of ethical studies for a number of decades.

ECO-CENTRED ETHICS AND DEEP ECOLOGY

Aldo Leopold, an American ecologist, writing as far back as 1949, tried to work out an eco-centred land ethic. He insisted that no progress could be made towards shaping such an ethic

---

12. The Report of the Working Party of the Catholic Bishops' Joint Committee on Bioethical Issues, *Genetic Intervention in Human Subjects*, London, Linacre Centre, 1996, p. 10.
13. Dieter T. Hessel, 'Now that Animals Can Be Genetically Engineered', *Ecotheology*, New York, Orbis Books, 1995, p. 285.

until the concept of land is expanded beyond the legal and economic domains. According to Leopold, looked at ecologically and ethically, land is a community which 'includes soils, waters, plants, and animals, or collectively: the land'.[14]

Leopold acknowledged that an ethic that might take such a concept seriously does not preclude using the land for human sustenance and welfare. It does, however, mean that land, in this sense, has a right to continue in existence in some way in the natural state. Leopold formulated his eco-centred principle as follows:

> A thing is right when it tends to preserve the integrity, stability and beauty of the biotic community. It is wrong when it tends otherwise.[15]

Those associated with the Deep Ecology movement would go further than Leopold in framing ethical norms that regulate human interaction with the rest of nature. Their focus is often called the ecocentric or biocentric approach because they argue that ethics should be concerned with the impact of human behaviour on ecosystems, like rivers, and even on the biosphere as a whole, as in the case of global warming. For Deep Ecologists, ecocentrism is both an ethical imperative and also a programme for political action. They insist that

> all things in the biosphere have an equal right to live and blossom and to reach their individual forms of unfolding and self-realisation within the larger Self-Realisation.[16]

Deep Ecology advocates, like the Norwegian philosopher Arne Naess, would not countenance the kind of experiments carried out on pigs or salmon, described above. Deep Ecology values non-human life independently of its usefulness to

---

14. Aldo Leopold, *A Sand County Almanac*, 1994 edition, New York, Ballantine Books, p. 239.
15. ibid p. 262.
16. Bill Davis and George Session, *Deep Ecology and Living as if Nature Mattered*, Salt Lake City, Gibbs-Smith, 1985, p. 64.

human beings. It is particularly keen on promoting social and ecological policies which involve non-interference with continuing evolution.

I would disagree with some positions espoused by Deep Ecologists, especially where they fail to acknowledge any unique dignity for human beings in the community of the living. Nonetheless, their insistence on the rights of other creatures and on the integrity of the ecosystem has brought an essential element into the debate about the ethics of contemporary biotechnology. For example, biotechnologists are now able to eliminate the brooding instinct in turkeys by blocking the gene that produces the prolactin hormone. Non-brooding turkeys are more productive than brooding birds. But is it right to engineer animals in a way that destroys their mothering instinct just to increase the profits of a particular business?

Fr Thomas Berry, an American priest who discusses ecological issues from a cosmological, ethical and religious perspective, also calls for pushing out the ethical boundaries to include all creation. He writes that contemporary ethics must focus its concerns on the larger community of the living. He argues that the human community is subordinate to the ecological community. The ecological imperative is not derived from human ethics. Human ethics is derived from the ecological imperative.

The basic ethical norm is the well-being of the comprehensive community, not just the well-being of the human community. The attainment of human well-being must be sought within the wider community.[17] Berry is well aware that the well-being of both humans and the earth are intimately linked. He is fond of saying that you cannot have healthy humans on a sick planet. What he wishes to emphasise is that the earth is a single ethical system, as is the universe.[18] For him this is the first principle of an ecological ethic, and any other ethical framework must be grounded on this bedrock.

17. Thomas Berry, *The Great Work*, Bell Tower, New York, p. 104.
18. Thomas Berry, 'Ethics and Ecology', unpublished paper 1994.

This position is at the opposite end of the moral spectrum from that which holds that humans are the only creatures on the planet that possess moral value. However, there are difficulties with Berry's position unless it is qualified by other important ethical principles. For example, in light of the extensive ecological devastation that humans have wreaked and are wreaking on the planet, one can see how humankind might be seen as a cancer of the biosphere. Some might argue that the rest of the earth's creatures would be better off without humans. An action that might rid the earth of this destructive creature could be seen, from this particular perspective, as something that was good for the planet and even for the continuation of evolution and biodiversity on the planet. That extreme articulation of this ecocentric position could give rise to a nasty form of ecofascism. As we will see below, an adequate environmental ethic has to keep a number of, sometimes competing, values in mind.

The more benign interpretation of Berry's position would call for a legal framework where the rights of the geological and biological as well as the human component of the earth community are clearly spelled out and protected. Obviously, the rights that we must accord to humans and the rights that we ought to grant to other creatures and entire ecosystems are not the same. The term 'rights' in this context is used in an analogous way. Berry is not conferring human rights, with the corresponding responsibilities, on animals; he is saying that animals and ecosystems have rights in and of themselves. Viewing the ethical choices from this perspective will at least help us to understand that the rest of the world is not simply there for humans to use or abuse.

The Australian theologian Denis Edwards argues that Christian ethical *praxis* is grounded in the understanding that creatures have ethical values independently of their usefulness to humankind. In other words, they have intrinsic value in themselves because of their relationship to the Creator

God.[19] Edwards makes the point that this is neither a ecocentric nor a geocentric ethical position but a theocentric one. Things have value in themselves because they are expressions of God's self-expression. 'Birds, plants, forests, mountains and galaxies have value in themselves because they exist and are held in being by the divine Persons-In-Mutual-Communion, and because they are fruitful expressions of divine Wisdom'.[20]

Edwards accepts that, until recently, the anthropocentric perspective has been very central to Christian ethical thinking. In the most acceptable articulation of a human-centred ethics, moral judgements and actions are based on upholding the dignity of each human person. In this tradition it is immoral to undermine or abuse the dignity of an individual human person even if one could argue that this is necessary for the welfare of the group.

Edwards goes beyond this human-centred approach when he states that 'the human person needs to be understood within a community of creatures, which have their own intrinsic value'.[21] He believes that there is no contradiction between affirming the intrinsic value of all creatures and respecting the unique dignity of each human person. In fact, this approach marries concerns for environmental well-being and for social justice.

Like Aldo Leopold, Edwards moves from considering individuals – humans or other creatures – to considering the biotic community.

> A Christian ecological praxis recognizes the interdependence of living creatures and gives particular ethical weight to biological communities, from local ecosystems to the biological community of the Earth.[22]

19. Denis Edwards, *Jesus the Wisdom of God*, Orbis Books, 1995, p. 154.
20. Ibid., p. 155.
21. Ibid., p. 156.
22. Ibid., p. 160.

This moral precept challenges individuals and communities to weigh up how an individual or collective action is to affect the community of life, now and in the future. Once again, we are caught up in a cyclic movement. If an action, like discharging nuclear or toxic waste in an inappropriate place, harms the community of life it will also harm the lives of individual humans and other creatures now and into the future. The fate of the individual is dependent on the welfare of the whole biotic community.

How, then, can one deal with competing claims? For example, when we study the story of the universe we learn that, at a particular moment in that story, life began to feed on life. This was an extraordinary breakthrough in the history of life and gave rise to a variety of ecosystems. However, only plants have the ability to take energy directly from the sun through the process of photosynthesis. All animals depend for their sustenance on plant matter or on eating other creatures that also feed on plants.

Therefore, Edwards' affirmation that all creatures have intrinsic value does not mean that he considers all to have equal value. He knows that ethical choices must be made between competing interests often involving humans and other creatures. Edwards weighs up the competing rights of cockroaches to live in his kitchen and his right, as a human being, to live in a hygienic environment. In resolving the conflict, he affirms that, since cockroaches and humans have intrinsic value, humans ought not arbitrarily destroy cockroaches. But because humans have a unique value in his moral calculus compared to cockroaches, humans can justify morally actions that expel cockroaches from their kitchens. Though the question of extinction of species is not addressed directly by Edwards, I presume that if cockroaches were facing extinction and this was their last habitat on the planet, this fact would tip the scales in their favour and humans would be morally obliged to live elsewhere. He does concede that 'there are times when [human needs] should take second place to the needs of the whole biological community'.[23]

Edwards is not convinced by Charles Birch's argument that intrinsic value is only conferred on creatures that have feelings, or by Peter Singer's and other animal rights proponents who argue that the ability to experience pain is the essential element in the moral argument. He insists that a theocentric ecological ethic attributes 'value to creatures not because they have feeling (Birch and Cobb) or because they experience pain (Peter Singer) but because they are the divine self-expression'.[24] Edwards believes that there is an ethical imperative on the human community to create a sustainable future for humans and the rest of creation. This may, for example, involve limiting the growth of the human population by acceptable forms of family planning.

In Edward's view the ethical imperative for ecological *praxis* is derived not from a single source of human-centred ethics nor even from focusing on the human and wider earth community. The sources that shape an acceptable environmental ethic include 'the intrinsic value of creatures, the dignity of the human person, the level of consciousness of a creature, the ecological whole and sustainability'.[25] Celia Deane-Drummond would add that we must approach nature with an attitude of humility, especially in light of the enormous and growing power of contemporary human technology to disrupt the geological and biological processes of the planet. She would like to see humans abandoning the *hubris* that seduces us into thinking that we can competently manage nature. Humility would move us from the high stool of resources management and counsel us instead to interact with nature in a more creative and less damaging way.[26] She feels that a theocentic ethical perspective which she shares

23. Ibid., p. 167.
24. Ibid., p. 162.
25. Ibid., p. 167.
26. Celia Deane-Drummond, op. cit., p. 72.

with Edwards would continue to stress the dignity of humankind: 'relativizing all human achievements from the perspective of a God who loves all creation, can give us a guide through some of the dilemmas'.[27]

Edward's moral principle of balancing competing moral claims with the wide framework of sustainability would prohibit the creation of the GE salmon and other animals that are being vigorously promoted by the biotechnology industry today. The right which animals have to preserve their own genetic integrity and the distress caused to them by GE technology would forbid such experiments.

From an ethical perspective, the nub of the issue revolves around whether other creatures have intrinsic value or not. If they do, then they have the right that their own 'specialness', especially the species boundary, be respected by humans. If they do not, and are merely objects, then of course there is no ethical imperative to respect their species uniqueness.

Even moral theologians who work exclusively within the Judeo-Christian tradition are beginning to insist on the intrinsic value of other creatures. In the tradition, as we have seen, animals and plants were considered to have value merely because they are perceived to be useful to human beings, rather than because they possess intrinsic worth in themselves. Starting from the position that other creatures do have intrinsic value, Professor James A. Nash would be very skeptical about the morality of genetically engineering other creatures. He writes that, since in the Christian tradition other species are deemed to have intrinsic value, the creation of transgenetic species should 'not be the norm but the rare exception on which the burden of proof rests. The genetic reconstruction of some species may be justified for compelling human needs in medicine, agriculture or ecological repairs (e.g. oil-eating microbes), so long as it can be reasonably tested and verified that tolerable alternatives are not available, genetic diversity

---

27. Ibid. p. 97.

is not compromised and ecosystemic integrity is not endangered'.[28]

Having assessed genetic engineering from an ecocentic and theocentric perspective, we can now turn to the question of potential impact on humans and the environment. The question is: Does genetic engineering pose such a threat to human health and the environment that the deliberate release of genetically engineered organisms should not be allowed at this point in time?

It is worth repeating the distinction between genetic engineering and other technologies. If a chemical and mechanical invention proves dangerous, it can be recalled or eliminated. This is not so with genetically engineered organisms. Even if only 1 percent of these organisms wreaks havoc on the environment, the consequences could be significant and irreversible because the organism will continue to reproduce and thrive. Even before the advent of genetic engineering, exotic species that have been introduced into an ecosystem have wiped out indigenous species and interfered with the complex web of relationships that exist between organisms in a given environment.

A good example of this is the story of what happened to the African killer bee. A noted Brazilian geneticist, Warwick Kerr, was experimenting with the bees when they accidentally escaped into the wild. This happened in the late 1950s. The bees have now spread throughout South and Central America and are moving north into the United States with devastating results for the environment and people.[29]

---

28. Quoted by Dierter T. Hessel, op. cit. Original quotation in James A. Nash, *Loving Nature: Ecological integrity and Christian Responsibility*, Abingdon, Nashville, 1991, pp. 61-2.
29. William Reville, 'Killer Bees from Africa Cause Havoc in South America', *The Irish Times*, September 1, 1997, p. 4.

OPPOSITION TO GENETIC ENGINEERING IN THE THIRD WORLD

Third World people are aware of the potential damage that genetically engineered organisms could do to their society and environment. Fifty peasant, indigenous and environmental groups from all over Latin America who gathered in Quito, Equador, in January 1999 to review developments in agricultural biotechnology, published a *Latin American Declaration on Transgenic Organisms*. The document rejects genetic engineering as being an ethically questionable technology which violates the integrity of human life and of species which have inhabited our planet for millions of years. It also points to the economic, social and environmental impact of genetic engineering as an unnecessary technology driven by commercial interest. The authors fear that if they adopt this technology they will become dependent on TNCs for food production, one of the most important functions in a subsistence community. They claim that money should be given to develop traditional and alternative agricultural technologies which do not pose such risks and which are compatible with the conservation of biodiversity.

In India, environmentalists, farmers and peasant groups have protested against the permission, given by the Indian Government to Monsanto, to field-test genetically engineered cotton seeds. In the Southern Indian States of Karnataka and Andhra Pradesh, farmers have begun to burn such fields in a campaign labelled 'Cremation Monsanto'. One of the main criticisms of the group is that Monsanto has not taken the biosafety measures they would be forced to take in trials in First World countries. They claim, for example, that there are no buffer zones between the test fields and the rest of the farm lands. Locals claim that in Europe the company would not be able to get away with such a policy.[30]

30. Gauri Lankesh and Pallavi Ghosh, "Indian Farmers Burn Genetically-engineered Crops', *Third World Resurgence*, December 1998/January 1999, pp. 2-4.

OPPOSITION FROM SCIENTISTS

By September 2000 over four hundred scientists, including people like David Bellamy, David Susuki and Vandana Shiva had signed an *Open Letter from World Scientists to All Governments*.[31] This letter calls for an immediate suspension of all environmental release of GE crops and products, both commercially and in open field trials for at least five years. They also insist that patents on all living processes, organisms, seeds, cell lines and genes be revoked and banned for the future. Finally, they call for a comprehensive public enquiry into the future of agriculture and food security.[32]

Professor Bevan Moseley, molecular geneticist and current Chair of the Working Group on Novel Food in the European Scientific Committee on Food, has drawn attention to the unforseen effect inherent in genetic engineering technology. He argues that the next generation of GE foods, such as vitamin A-enriched rice, will be even more dangerous because of the complexity of the gene construct.[33]

The Japanese Canadia professor, David Suzuki, who has been researching and teaching genetics since 1961, smiles when he reflects on how the certainties that he held in the 1960s have all vanished. He wrote:

> Today when I tell students the hottest ideas we had in 1961 about chromosome structure and genetic regulations, they gasp and laugh in disbelief. In 1997, most of the best ideas of 1961 can be seen for what they are – wrong, irrelevant or unimportant ... So what is our hurry in biotechnology to patent ideas and rush products to market when the chances are overwhelmingly that their theoretical rationale will be wrong?[34]

31. 'Open Letter from World Scientists to All Governments concerning Genetically Modified Seeds', *Institute of Science in Society*, www.i-sis.org.uk/ May 17, 2003, p. 4.
32. Ibid p. 2
33. Ibid p. 5.
34. David, Suzuki, 'Can Science "Manage" Nature?', *The Ecologist*, January/ February 1998.

Closer to Ireland, another well-known environmentalist, David Bellamy, acknowledges that 'genetically modified products worry him'.[35]

The British Medical Association, in their interim report published in 1999, called for an indefinite moratorium on the release of GE organisms until there are further studies on new allergies, the spread of antibiotic-resistant genes and the effects of transgenic DNA.[36]

The Union of Concerned Scientists (UCS) in Washington, in their 1993 report on genetic engineering entitled *Perils amidst the Promise*, also promote the idea of a moratorium. They concluded that no company should be permitted to commercialise a transgenetic crop in the USA until a strong government programme is in place that assures risk-assessment and control of all transgenic crops. On biodiversity, they call for adequate protection for the centres of crop diversity in the US and elsewhere in the world.

In 2003, the UCS position is still cautious. While not supporting or opposing genetic engineering as such, it finds that the risk-benefit calculus, especially with regard to GE food, is complicated. It believes that there is little certainty about either the risks or benefits of GE food and, in the absence of such knowledge, it is perturbed about the rapid commercialisation of GE food. The UCS has called on the US government to strengthen the regulatory regime governing genetically engineered organisms – plants, animals and micro-organisms – so that the risks and benefits can be properly evaluated on a case-by-case basis before any food is allowed to enter the food chain. UCS scientists are afraid that engineered organisms might harm people's health or the environment. They might kill beneficial insects and jeopardize valuable natural resources like Bt toxins. The UCS has also called for much more informed public debate about genetic engineer-

35. Roisin Ingle, 'Bellamy Happy to be a Bogman', *The Irish Times*, 4 April 1998, p. 10.
36. Ibid p. 6.

ing and the direction in which it is now being taken by biotech corporations.[37]

OPPOSITION FROM RELIGIOUS LEADERS

Pope John Paul II in a talk to an estimated fifty thousand Italian farmers on November 12, 2002 exhorted them to 'resist the temptation of high productivity and profit that work to the detriment of the respect of nature'. The Pope added: 'when [farmers] forget this basic principle and become tyrants of the earth rather than its custodians ... sooner or later the earth rebels.' Later in the talk he returned to this theme and said that if modern farming techniques do not 'reconcile themselves with the simple language of nature in a healthy balance, the life of man will run ever greater risks, of which we already are seeing worrying signs.' While the Pope did not specify what these risks might be, commentators believed he was speaking about the risks involved in genetically engineered foods.[38]

The Pope returned to the idea of life as a gift and the limitations of human intervention in his Lenten Letter of 2002, *You Received without Paying, Give without Pay*.

> May these words of the Gospel echo in the heart of all Christian communities on their penitential pilgrimage to Easter. May Lent, recalling the mystery of the Lord's Death and Resurrection, lead all Christians to marvel in their heart of hearts at the greatness of such a gift. Yes! We have received without pay. Is not our entire life marked by God's kindness? The beginning of life and its marvelous development: this is a gift. And because it is gift, life can never be regarded as a possession or as private property, even if the

---

37. 'USC's Position on Biotechnology', Union of Concerned Scientists, www.ucsusa.org/food_and_environment/biotechnology/ p. 1, May 29, 2003
38. Pope John Paul II, Talk to Italian Farmers, reported by Eric Lyman for The Bureau of National Affairs, Inc. Washington D.C. November 15, 2000.

capabilities we now have to improve the quality of life can lead us to think that man is the 'master' of life.

The achievements of medicine and biotechnology can sometimes lead man to think of himself as his own creator, and to succumb to the temptation of tampering with 'the tree of life' (Genesis 3:24).

It is also worth repeating here that not everything that is technically possible is morally acceptable. Scientific work aimed at securing a quality of life more in keeping with human dignity is admirable, but it must never be forgotten that human life is a gift, and that it remains precious even when marked by suffering and limitations. A gift to be accepted and to be loved at all times: received without pay and to be placed without pay at the service of others.

In November 2000, the South African Catholic Bishops' Conference expressed its concerns about genetic engineering in agriculture and food production. The Archbishop of Durban, Wilfred Cardinal Napier, said that GE is an imprecise technology and that the long-term health effects of consuming GE food have not be fully assessed. Speaking about irreversible damage to the environment, he pointed out: 'Because we do not know whether there are any serious risks to human health or the environment, to produce and market genetically modified food is morally irresponsible. The precautionary principle should apply, as it does in medical research.' [39]

The Archbishop of Manila, Jaime Cardinal Sin, made much the same point in a pastoral statement on genetic engineering in agricultural products published on May 8, 2001. He stated: 'genetic engineering is acceptable only if all risks are minimized. Otherwise, one may easily succumb to temptations of productivity and profit at the expense of the

---

39. 'Church Calls for Moratorium on Genetically Engineered Food', www.allafrical.com/stories.2001, p. 1.

people and environment. And as long as foreseeable dangers are not fully identified, studied and avoided, safe alternative procedures should be used'.[40]

In the Philippines, the Bishop of Marbel, Dinualdo Gutierrez, condemned the field trial of Bt corn in South Cotabato in June 2001. Further north in Luzon, Bishop Sergio Utleg of the Diocese of Ilagan wrote a pastoral letter against field trials for genetically engineered crops. After stating that 'the Church is not against development and modern technology', he went on to ask questions: For whom is this development? Who is to benefit from these so-called development projects and modern technology? According to the bishop, development must serve the needs and promote the progress of all people.[41]

Even though the Catholic Church has attempted to develop a creation theology in recent years and in some pronouncements has questioned genetic engineering of plants and animals, Rome does not seem to have a consistent position on this question. We saw earlier that Bishop Elio Sgreccia, vice-president of the Pontifical Academy for Life, believes that there are no impediments to animal and vegetable biotechnologies.[42]

In August 2003, Archbishop Martino, head of the Pontifical Council for Justice and Peace, said that the Vatican was preparing an official report on biotechnology which would come down in favour of genetically modified foods. The reason he gave was that GE foods would help alleviate starvation, a position which, as I have shown in Chapter Four, is open to challenge. The Archbishop went on to state that he had lived in the US for sixteen years and that he ate everything that was offered to him including genetically engineered

40. Pastoral Statement on Genetic Engineering, read at Masses on May 8. 2001.
41. Aquiles Z. Zonio, 'Church Urges Gloria [President Macapagal-Arroyo] to Stop Bt-corn Tests', *Inquirer*, November 23, 2001, p. A
42. www. zenit.org, Rome, October 9, 2002, p. 2.

food, with no ill-effect on his health.[43]

The Archbishop's comment raises questions about his knowledge of genetically engineered foods. Because GE soybeans have been available extensively only since 1997 he was not eating GE foods for most of his time in the US. Furthermore, as we have seen repeatedly in Chapters Four and Five, the scientific jury is out on whether GE foods damage human health. Small farmers in Third World countries and their spiritual leaders like Bishop Dinualdo Gutierrez of Marbel in Mindanao will feel that they have been abandoned in favour of biotech corporations who are poised to make billions of dollars on GE seeds.

OPPOSITION TO GE CROPS IS MORE RESOLUTE IN EUROPE

Opposition to GE foods has been much more vigorous in Europe than in the United States. In the last ten years the European consumer has become much more concerned about food safety in the wake of the BSE and salmonella scares. European consumers were not always convinced about the way their governments coped with these incidents. Europeans are also more cynical about corporate aims, and gastronomy is central to the cultures of many European countries.

The biotech industry, especially Monsanto, misjudged the reaction of European consumers over genetically engineered products. They thought that they could introduce transgenic products with little hostile reaction – as they had previously done in the US. They were surprised at the opposition they encountered in Europe.

A study in Germany in 1998 found that only 15 percent favoured GE food while 81 percent were opposed to it. In April 1998, Philip Angell, Monsanto's US director of corporate communications, stated: 'we made a mistake which we regret. We should have listened more carefully.' Environmen-

---

43. Richard Ower, 'Vatican Backs "Life-saver" GM Crops', *The Irish Independent*, August 4, 2003, p. 22.

tal organisations believe that this remorse is not genuine. They see the company's new willingness to listen as a pretence and its new-found openness to dialogue with the opposition as merely another ploy to speed up acceptance of genetically engineered products by consumers.[44]

In October 1998, the *New Scientist* reported that other biotech companies were very critical of the tactics used by Monsanto in Europe. These companies feel that Monsanto is 'largely to blame for a consumer backlash that could cripple the prospects for genetically engineered food in Europe ... A high-profile advertising campaign from Monsanto, designed to reassure European consumers, has, if anything, hardened negative public attitudes to agricultural biotechnology. "We are as fed up as some others with the Yankee-Doodle language that comes to our consumers," says Greef of Novartis'.[45] The Advertising Standards Authority (ASA) in Britain condemned Monsanto's advertising campaign for making claims about GE foods that were 'confusing, misleading, unproven and wrong'. Claims that GE crops were grown in environmentally sustainable ways were also dismissed, and the suggestion that Monsanto would sacrifice sales of its herbicide Roundup Ready to reduce pesticide use when it had no intention of doing so, was 'confusing'.[46]

The trenchant opposition to GE crops in Europe, especially in Britain, meant that many supermarket chains were forced by consumer pressure to source their own brands from places that did not plant transgenic soya or maize. The supermarket chain Iceland, with 770 stores in Britain and 6 in Ireland, decided to ban all foods containing genetically engineered organisms. The founder, Malcolm Walker, who at the time was also chief executive, accused biotechnology compa-

44. John Vidal, 'Food Firm's PR Errors', *The Guardian*, April 13, 1998.
45. Andy Coghlan, "Mutiny against Monsanto', *The New Scientist*, October 31, 1998, pp. 3 and 4.
46. Sarah Hall, 'Monsanto Ads Condemned', *The Guardian*, March 1, 1999, p. 5.

nies of 'conning' Irish and British consumers, and claimed that genetically engineered food is being introduced 'by stealth'. Iceland claimed its stance on GE foods was based on ethical principles. Health Food shops in Britain have also taken transgenic products or ingredients off their shelves. The policy will include sauce mixes and vegetarian burger mixes.[47]

Sainsbury's was the first supermarket to stock GE foods when it sold GE-produced tomato puree in February 1995. Other supermarkets like Marks and Spencer (M&S) were initially favourable to GE food. M&S assured their customers that genetic modification could have the potential to offer customers direct benefits in new products which should be assessed on their own merits.[48] The company and other supermarket chains like Sainsbury's were put under pressure from consumers and elements in the media to reverse its policy. In March 1999, M&S announced that it was banning all GE food and food ingredients. It told its customers: 'we will be the only major retailer where customers can purchase any product on the shelves with full confidence that no GE ingredients or their derivatives have been used'.[49] In fact, Asda and other supermarket chains had taken a similar option and were working with suppliers to ensure that their own-brand products do not contain GE soya or corn. This increase in demand led to a marked increase in the price of non-GE corn and maize in 2001 and 2002.

The Irish supermarket chain Superquinn published a pamphlet on genetically modified food early in 1999. It accepted the biotech industry's argument that there is a direct continuity between food modification using either natural or biotechnological means. It proclaims that 'the benefits from

---

47. James Meikie, 'Gene-modified Products Barred from Health Food Stores', *The Guardian*, June 24, 1998.
48. *Selling Suicide: Farming, False Promises and Genetic Engineering in Developing Countries*, Christian Aid, PO Box 100, London SE1 7RT, p. 27.
49. Ibid.

genetic modification are manifold and include disease resistant crops; crops that require less herbicides and pesticides in their production; longer lasting fruits and vegetables, and foods with higher vitamin, mineral and protein contents and lower fat contents.' The document admitted that there are concerns but it did not list them in the way it lists the so-called benefits. Superquinn promised that it would label foods that are 'genetically modified or produced from genetically modified soya and genetically modified maize.' However, they would not label 'oils or other soya and maize derivatives ... as they do not contain certain modified protein.' They claimed that the oils are identical to the oils from the non-modified seeds and do not contain genetic material. This is another version of the 'substantial equivalence' argument.

Mr John McKenna, a food writer with the *Irish Times*, told the Irish Association of Health Stores that consumers do not want genetically modified organisms in their food. 'There is no demand from any quarter other than the producers of GMO food'.[50]

A survey on Consumer Attitudes to Genetic Engineering and Food Safety, commissioned by the environmental organisation, Genetic Concern, was carried out in Ireland by Lansdowne Market Research in January 1999. It found that very few people felt that they were well informed about genetic engineering. 78 per cent said that they knew little or nothing about the technology. Most of those who felt they were well informed, 89 percent, were concerned about the implications of genetic engineering for food safety. Only 8 per cent were unconcerned. Generally speaking, women, especially married women, were more concerned than men. This is hardly surprising since women do the bulk of household weekly shopping. When asked whether they were concerned about genetic engineering, a clear majority of those interviewed said they were.

50. *The Irish Times*, October 20, 1997.

THE PRECAUTIONARY PRINCIPLE

Given the dangers associated with genetically engineered organisms outlined in Chapter Four and the lax regulatory climate portrayed in Chapter Five, given also the opposition from Third World farmers, from scientists, religious leaders and consumers, I believe that there is an excellent moral case for invoking the precautionary principle.

This would call for instituting a five-year moratorium on the deliberate release of genetically engineered organisms until the risks are much more clearly understood. Five years would also give enough time for a thorough public discussion about all the issues involved from a variety of perspectives – health, ecological, ethical and religious. The moratorium would include a ban on the commercial growing of GE crops, the importation of GE food and ingredients, and their use in human and animal feeds.

One articulation of the precautionary principle emerged from a meeting of activists, scholars, scientists and lawyers, at Wingspread, home of the Johnson Foundation in Racine, Wisconsin, in 1998. The group was convened by the Science and Environmental Health Network (SEHN). The Wingspread definition of the principle contains three important elements: the threat of harm; scientific uncertainty; and preventative, precautionary action. It reads as follows:

> When an activity raises threats of harm to human health or the environment, precautionary measures should be taken even if some cause and effect relationships are not fully established scientifically. In this context the proponent of an activity rather than the opponents should bear the burden of proof. The process of applying the precautionary principle must be open, informed and democratic and must include potentially affected parties. It must also involve an examination of the full range of alternatives, including no action at all.[51]

51. www.johuston/dn.org.conference/precautionary/, p. 2.

In the wake of the controversy surrounding Professor Pusztai's findings that rats fed with GE potatoes developed cancerous tumours, forty environmental and religious organisations in Britain called for a five-year ban on GE food. This call for caution seems eminently reasonable given the fact that the issues are so grave and the consequences of a mistake are so serious. As humans have learned from bitter experience, it is impossible to recall a living organism.

COUNTRIES LIKE IRELAND AND THE PHILIPPINES SHOULD FOLLOW NORWAY'S EXAMPLE

In 1996, Norway adopted a consultation model on biotechnology which was developed by the Danish Board of Science. In the process, groups of ordinary people assessed the various aspects of biotechnology, including the ethical, economic, political, social and legal perspectives, in addition to the narrower technological considerations, before deciding whether Norway should opt for biotechnology. The panels concluded that there was no need for genetically modified food in Norway today, because the selection, availability and quality of ordinary food is satisfactory. Too many uncertain factors attach to genetic engineering.

Ireland would do well to follow this example until much more is known about genetically engineered organisms. Unfortunately, the consultation process adopted by the then Minister for the Environment, Noel Dempsey, in 1999 was totally unsatisfactory from the perspective of the participating NGOs. They felt that the debate with the biotech industry was superficial and very poorly structured and amounted to little more than an exchange of sound-bites. For this reason, the majority of NGOs withdrew from the process. It came as no surprise to the NGO community that the consultation document, published in October 1999, came down on the side of the biotech industry. The report misrepresents the biocentric ethical argument when, on page 22, it states that it tries to set the well-being of the comprehensive community over against the well-being of the human community.[52] What is needed is a much more transparent and participative process. To date, both of these facets have been singularly absent.

Unfortunately, during much of the last three years there is far less debate on genetic engineering. One of the major reasons for this is that the Irish NGO, Genetic Concern, collapsed in 2000. The handful of people who ran the organisation just could not continue to keep going under pressure. This is good news for the corporate world.

Another set-back to a GE-free world happened, this time in the Philippines, on December 5, 2002. Monsanto announced that their GE Bt corn variety called 'Yieldgard' had been approved for commercial release by the Philippine Department of Agriculture. The potential for polluting non-Bt corn is enormous for two reasons. In the Philippines, the average size of a farm is only 1.5 hectares; and, secondly, corn is wind-pollinated. So, even if the Department of Agriculture in the Philippines demands that there be a significant distance between GE and non-GE corn, it will only be a matter of time before all the corn grown in the Philippines is either GE corn or GE-contaminated corn. Thus will be achieved the aim of the corporations to completely control the corn seed markets. In the Philippines, Monsanto already control 30 percent of the commercial corn seed market, while DuPont/Pioneer control a further 60 percent. The decision to allow the commercial planting of Bt corn will have an enormous negative impact on the lives of the 600,000 corn farmers in the Philippines. It is a death knell especially for farmers with small holdings.

Critics of the decision point out that the field-test tested only the effectiveness of the Bt protein against the corn borer. No tests were undertaken on the impact of widespread dis-

---

52. *National Consultation Debate on Genetically Modified Organisms and the Environment*, Department of the Environment, Dublin, July 28,1999.

semination of Bt corn on human health and the environment. There were no tests on how quickly pests adapt to the Bt technology. If insects adapt within two or three years, farmers will be forced into a 'genetic threadmill' of constantly needing new genetically engineered fixes to deal with insects that have developed a resistance to Bt.

At a one-day meeting of the Catholic Bishops Conference of the Philippines (CBCP), in February 2002, the bishops called on President Macapagal Arroyo to defer the widespread planting of Bt corn until more research on its impact had been carried out.

The environmental organisation Greenpeace called on the Philippine Government to revoke their decision to allow the commercial planting of GE Bt corn. They also called for genuine public participation and transparency on all decisions regarding the introduction of genetically engineered organism.[53]

Eighteen doctors and other medical professionals from the University of the Philippines wrote a letter expressing their strong opposition to the government's decision to approve the commercialisation of Monsanto's Bt corn. The Bureau of Plant Industry (BPI) and all those responsible for allowing Bt corn to be openly marketed in the country have irresponsibly ignored the potentially serious consequences of the widespread dissemination of genetically modified organisms (GMOs), particularly Bt corn. In doing so, the government has practically reneged on its constitutional mandate to protect the people's right to health and a healthy environment.

Since December 2002, Philippines environmentalists and small farmers have been protesting the decision to allow Monsanto to sell its Bt corn. When the Department of Agriculture refused to enter into meaningful dialogue with them on

---

53. Greenpeace briefing paper, December 13, 2002, available on their website: www.greenpeace.org

this crucially important issue, a group of Filipino environmentalists decided to go on hunger strike. They began their hunger strike in front of the Department of Agriculture (DA) on April 22, 2003, and continued for twenty-nine days. Their action captured the imagination of many Filipinos. Despite a surfeit of news stories, global and local, like the Iraq war, Severe Acute Respiratory Syndrome (SARS) and the conflict in Mindanao, the witness of the hunger strikers managed to keep the Bt corn issue before the Philippine public like no other event in recent Philippine history. The protesters broke their strike on the thirtieth day because of failing health. They have now committed themselves to fighting against GE crops in the rural areas of the Philippines. They warn farmers against planting what they call these 'poisonous' seeds.

In this chapter on ethics and genetic engineering I have attempted to show that the moral framework which has emerged from the Western tradition is not sufficient to deal with questions raised by genetic engineering. This tradition is excessively human-centred and has given little thought to the ethical status of other creatures, ecosystems or the biosphere. I argued that, because there are so many scientific, social and developmental problems surrounding genetic engineering, there should be a five-year global moratorium on the direct release of genetically engineered organisms into the environment. Having looked at the data the precautionary principle would dictate such a course of action.

SEVEN

Stop Patenting Life

IT IS IMPORTANT, at the beginning of this chapter on patenting, to make a clear distinction between the morality of genetic engineering and the morality of patenting living organisms. In Chapter Six I argued that an ethical evaluation of genetic engineering in 2003, especially as it affected agriculture, warranted a call for a five-year moratorium until the current problems with the technology and the justice and equity issues are sorted out. The biologist and theologian Celia Deane-Drummond, while evaluating the danger which genetic engineering poses to human health, to the environment and to the task of building a more equal and just world order, makes a good case for not ruling out genetic engineering in principle.[1] She writes: 'a theological approach encourages those who are involved to see the wider social and religious implications of these decisions. It does not necessarily ban all genetic engineering, but seeks to transform it so that it more fully represents a humane enterprise'.[2]

However, I believe there is an overwhelming moral case against patenting life. Over time, the patenting scramble will remove many life forms from common, shared ownership under which they have provided many services for humans and other creatures. Under a patenting regime, these life forms will become the private property of Northern transnational corporations. Life will have value only in so far as it generates a profitable return on investment for large companies. The debate is timely because the issues were on the agenda of the World Trade Organisation when it met in Cancún in Mexico in September 2003. Many Third World coun-

1. Celia Deane-Drummond, *Theology and Biotechnology*, Geoffrey Chapman, London, pp. 80-104.
2. Ibid p. 99.

tries and non-government organisations (NGOs) wanted to change radically Article 27.3(b) of the Agreement on Trade-Related Aspects of Intellectual Property Rights (TRIPs). These groups fear that a global regime of patenting will fill the coffers of rich Northern transnational corporations and further impoverish the poor, especially in the Third World.

Life, which was once considered sacred and a gift from God in almost all the religions and cultures of the world, is now seen as a human invention, a collection of genes and chemicals that can be engineered and bought and sold by a patent holder.

Such a reductionist, mechanistic and materialistic concept of life is at variance with the tenets of all major religions. With patenting, human beings claim to have invented plants and animals and to have exclusive control over them. If the scramble to patent living forms gathers pace across society, it will undoubtedly devalue the meaning of life. No part of the Earth will be sacred in the future. Furthermore, it could well mean that within a few decades 'the entire human genome ... [will] be owned by a handful of companies and governments'.[3]

PRIVATIZING THE COMMONS

The similarity between what happened with the Enclosure Acts in Britain in the eighteenth century, and what is happening today with global Trade-Related Intellectual Property legislation, has not been lost on commentators. Pat Roy Mooney points out:

> The rich landlords who orchestrated the enclosure movement ... argued that the commons must be privatized so that they could take advantage of the new agricultural technologies and feed growing urban populations ... In the same way and with the same arguments as the Enclo-

---

3. Andrew Kimbrell, *The Human Body Shop*, Harper, San Francisco, 1993, p. 190.

sure Acts used to drive rural societies from their ancestral lands (and rights), TNCs are now pursuing another Enclosure Act – the intellectual property ('IP') system – to privatise the intellectual commons and monopolise new technologies based on these commons. The Landlords have become the Mind Lords. In the post-GATT world of new biotechnologies, these are also the Life Lords.[4]

What is happening since the latter part of the twentieth century is a new, more invidious form of colonialism. The goal this time is not just to conquer new lands, as Vasco da Gama, Columbus, Magellan or Cromwell did, or to lay claim to gold or precious stones – it is to colonise life itself. Many of the agribusiness, pharmaceutical and biotech corporations involved in this enterprise are larger financial entities than the average nation state. They can bring enormous pressure to bear on politicians, both nationally and globally, to design a regulatory regime that promotes their products. Since most of the multinationals have their headquarters in the US, they have persuaded the US government to write to Third World countries warning them that unless they stop importing generic drugs the US will withhold special trading privileges.[5]

THE RATIONALE BEHIND PATENTING

The rationale behind patenting is the desire to reward and compensate an individual for the time and expense involved in developing an invention. The individual is normally granted monopoly rights over his/her invention for between seventeen and twenty years. The patent-holder can prevent other people from making, using or selling the invention unless they pay a licence fee or royalty on any commercial application derived from the invention.

Three criteria are required in order to obtain a patent for

---

4. Pat Roy Mooney, 'Private Parts: Privatisation and the Life Industry', *Development Dialogue*, April 1998, p. 138.
5. Charlotte Denny, 'Patently Absurd', *The Guardian*, April 9, 2001.

an invention which can be either a material product or a process. It must be new or novel; it must involve a non-obvious inventive step; and, finally, it must be useful and have a commercial application.

It is worth noting that the approach of biotech companies to regulatory agencies and patent offices are inconsistent. They tell the regulatory agencies that genetic engineering is not new; it is just an extension of traditional breeding procedures. Because of this statement, GE crops are not treated as a novel food and are not subjected to rigorous testing. When biotech companies arrive at the patent office, however, they sing a different tune. Here, they claim that genetic engineering is new and that they should be allowed to patent their new creation.

The truth is that the geneticist or biotechnologist does not create *de novo* genes, cells or organisms. They identify, isolate and modify these entities, which is a very different operation from creating them. Many people would suggest that the analogy between a chemist patenting the elements of the periodic table and a geneticist patenting genes is appropriate. Jeremy Rifkin states: 'no reasonable person would dare suggest that a scientist who isolated, classified and described the properties of hydrogen, helium or oxygen ought to be granted the exclusive right, for twenty years, to claim the substance as a human invention'.[6] For this reason patents should not be given for living organisms. Other mechanisms ought to be developed to protect the legitimate financial interests of those who invest in biotechnology products or procedures.

PATENT LAWS FRAMED FOR INDUSTRIAL PRODUCTS AND VARY FROM COUNTRY TO COUNTRY

It is important to remember that patent laws were framed in an industrial context and therefore are more suitable for machines than for nowledge. The first recorded patent was

---

6. Jeremy Rifkin, *The Biotech Century*, Victor Gollancz, London, 1998, p. 45.

given to the architect and engineer, Filippo Brunelleschi, in 1474 for a machine that would lift marble on to a barge, and it was for three years. Brunelleschi's main legacy is the beautiful dome of the cathedral in Florence. Though a patent law appears in Britain in 1623, patenting did not really come into force in that country until 1852. In the United States patents were granted on imported technologies without any proof of originality.[7] Patents have been sought for objects, chemicals, designs and processes.

Until recent times, patent laws differed from country to country, reflecting the way in which different cultures and political systems weighed up the often conflicting claims between compensating the inventor and ensuring that the public would benefit from the new product. The pendulum normally tilted in favour of the common good of the nation rather than towards securing the financial interest of the inventor or the corporation. Most Third World countries, for example, refused to recognise patents on food and medicine and other basic products that are deemed basic human needs. When Alexander Fleming invented penicillin at St Mary's Hospital in London in 1928, the British government decided that this drug should not be patented because of the potential value to humankind.[8]

Earlier patent agreements began with the Vienna Congress in 1873. This was followed by the Paris Convention of the International Union for the Protection of Industrial Properties. Signed initially by eleven countries, it was revised in 1911, 1925, 1934, 1958 and 1967. The Berne Convention on copyright, signed in 1886, was updated in 1946. This Convention recognised that individual countries had particular needs and priorities and that these would be reflected in national patent legislation.

It is worth remembering that in many industrialised coun-

7. Ha-Joon Chang, *Kicking Away the Ladder*, Anthem Press, P.O. Box 9779, London, SW19 7QA, 2002, pp. 83-85.
8. Denny, art. cit., p. 9.

tries like France, Germany, Japan, Switzerland, Italy and Sweden, patenting legislation appeared only after the development of their own industries and that, even after signing the international agreements on patents, some countries seldom enforced them.[9] The development of the textile industry in the United States in the early nineteenth century was based on patterns and machines which were developed in Lancashire. The Japanese textile industry followed this same route in the early twentieth century, and that country's much-vaunted economic miracle in the post-World War II period was based on innovative copying. At the end of the nineteenth century Germany complained about the absence of a patent law in Switzerland and the consequent theft of German intellectual property by Swiss firms, especially in the chemical industry.[10]

The first break with these country-specific patent laws took place during the Uruguay Round of the General Agreement on Tariffs and Trade (GATT) which was concluded in 1994. Under pressure from its corporate sector, the US, together with other Northern countries, pushed for 'harmonisation' in the laws on intellectual property rights across the world. It is worth noting that 70 per cent of US export earnings are linked to patented items, from AIDS drugs to Disney, McDonalds and Microsoft. The resulting GATT Agreement on Trade-Related Aspects of Intellectual Property Rights (TRIPs) obligated all GATT signatories to adopt minimum intellectual property standards for plants, animals, microorganisms and biological material, including genes.

Gradually, the understanding that patenting applied only to inanimate things and processes began to be eroded. It is no secret that the giant agribusiness Cargill was largely responsible for drafting the Agreement on Agriculture at the World

9. Eva Ombaka, 'Trade-Related Aspects of Intellectual Property Rights (TRIPS) and Pharmaceuticals', in *Echoes*, 15/1999, World Council of Churches, P.O. Box 2100, 1211 Geneva, Switzerland.
10. Chang, op. cit., p. 57.

Trade Organisation (WTO) – despite the fact that it is not even a public company. [11]

The biotech industry claims that patents are necessary so that innovative, life-saving technologies will be developed. Critics counter that this argument has no historical support. In fact, the opposite is the case. Until the middle of the nineteenth century Switzerland was an agricultural country, poor in natural resources. Because there was no patent law a small company copied the aniline dyeing process which had been developed and patented in Britain. That company, which later was called Ciba, developed into a major global enterprise. In 1995 it merged with another Swiss company called Sandoz to form Novartis. Ironically, Novartis led the campaign in Europe which allowed companies to patent genes and life.

Companies often call for patents to pay for innovation. But Eric Schiff, a historian of economics, shows that no country has contributed as many basic inventions as did Switzerland during her patentless period. These inventions include milk chocolate (Daniel Peter, 1875), chocolat fondant (Rudolf Lindt, 1879), and powdered soup (Julius Maggi, 1886). Holland followed a similar path. In 1870, two Dutch companies, Jurgens and Van Den Bergh, used a French patented recipe to produce margarine. These later merged with the British company Unilever – now vigorously promoting patent legislation. Similarly, Gerard Philips began manufacturing light bulbs – invented by Thomas Edison. Schiff argues that on economic grounds it is difficult to avoid the impression that the absence of patents 'furthered rather than hampered development'.[12]

The Cambridge economist Ha-Joon Chang, in his book,

11. Katharine Ainger, 'The New Peasants' Revolt', *The New Internationalist*, January/February 2003, p. 11.
12. George Monbiot, 'Companies now Demanding Intellectual Property Rights were Built up without Them', *The Guardian*, March 12, 2002, p. 15. (Quoting from Schiff, Eric: *Industrialization without National Patents*, Princeton University Press, 1971).

Kicking Away the Ladder: Development Strategies in Historical Perspective, makes it quite clear that history debunks the free-trade myth. He points out that countries like the United States, Switzerland and Holland became rich on the basis of protectionism and subsidies. Once rich, these countries began pressurising poor countries to accept so-called free-trade and all its accoutrements, including a restrictive patenting regime – policies which, Ha-Joon Chang maintains, have killed growth in many Third World countries, especially in Africa and Latin America. In order to stimulate growth the WTO ought to rewrite its rules 'so that developing countries can more actively use tariffs and subsidies for industrial development'.[13]

Furthermore, patents enable companies to create a monopoly on a product, permitting artificially high pricing. As a result, many products and procedures, including drugs, will be priced out of the reach of poor people. Third World critics of the Northern-dominated pharmaceutical industry point out that these corporations spend millions of dollars researching profitable lifestyle drugs like Viagra, but neglect the diseases of the poor, like malaria and tuberculosis, to mention just two.

The court case in South Africa where forty transnational pharmaceutical companies took the South African government to court to prevent it importing generic drugs which are needed in the fight against AIDS illustrates the determination of TNCs like GlaxoSmithKline to protect their patents at any cost. The usual rationale that the companies give for seeking patents – huge research and development costs – did not pertain in this case since the medication was developed in public institutions and has been leased to a pharmaceutical company. The disparity in costs was staggering. In 2001, ciprofloxican, which is an essential medicine for AIDS sufferers, costs South Africa's public health sector Stg£0.52 per pill

13. 'History Debunks Free Trade Myth', *The Guardian*, June 24, 2002, p. 23 (book review).

and the country's private health care providers more than £3 per pill. If the new law is implemented, a generic drug could be imported from India for Stg£0.04 per pill.[14] Obviously, access to generic drugs would be good news for the 37 million people suffering from AIDS in Africa alone.

The court action, which was followed with interest around the world, turned into a PR disaster for the giant pharmaceuticals. They appeared rapacious and greedy, willing to put their profits before the well-being of millions who are suffering from AIDS. This greed was seen again in December 2001 when the Competition Commissioner of the European Union, Mario Monti, fined a number of pharmaceutical and chemical companies €1.5 billion for price-fixing and acting as a cartel. The Swiss chemical company Hoffman-LaRoche was fined €462 million for conspiring to fix vitamin prices. The controlling power of transnational companies can be seen in the fact that this extraordinary scandal did not make the front page or top story in the media. I found the information in the *Irish Times* financial section, below the fold.[15]

The double-standard in the North's approach to patented medicine was once again revealed during the anthrax attacks in the US in October 2001. Fearing widespread anthrax attacks on the population of the US and Canada, the US considered breaking the patent on Cipro (an anthrax antidote), and the Canadians actually did break the patent, in order to produce sufficient quantities of the drug. At the Doha meeting of the WTO in 2001, the US delegation was forced to allow countries to buy generics drugs for AIDS patients at a fraction of the cost of the patented medicine. After all, the US government had threatened to overrule a patent on an anthrax medicine after twelve people in the US had died from anthrax. At Doha, negotiators from other

14. Chris McGreal, 'Crucial Drug Case Opens in Pretoria', *The Guardian*, March 6, 2001.
15. 'EU Sends Strong Message to Cartel Price-fixers', *The Irish Times*, December 6, 2001, p. 15

countries were asking: Are the lives of twelve US citizens more important than thirty million Africans? Since the Doha meeting, however, the US, after strong lobbying by American pharmaceutical giants, has refused to relax global patents on a full range of life-saving drugs.[16]

The US has been pushing the free-trade agenda because it benefits its transnational corporations. When their interests are threatened, the US becomes very protectionist. The 2002 Farm Bill gives subsidies in the region of $248.6 billion to corporate agriculture. This subsidy will have a very negative impact on Third World agriculture. The present administration has also protected US steel manufacturers.

As we saw above, the parent companies of some of the most pro-patenting companies in today's world were once against patenting. In the mid-1800s the parent company of what later became Ciba-Geigy Ltd. was fighting any attempt to establish patent laws in Switzerland. There is a modern ring to their arguments. They claimed that '[p]atent protection forms a stumbling block for the development of trade and industry... The patent system is a playground for plundering patent agents and lawyers.' [17]

The labyrinth of patenting legislation has affected one of the proudest of GE foods, golden rice. The biotech industry boasts that genetically engineered rice could help prevent blindness among poor children. Millions of dollars of public funding was spent developing this technology which was hailed as proof that biotechnology will help feed and supplement the diet of the poor who might be lacking in Vitamin A. The researchers, Ingo Potrykus and Peter Beyer, who developed the transgenic beta-carotone enhanced rice, were so afraid of the complexities of patent negotiations that they quickly signed the publicly-funded technology to AstraZeneca

16. Larry Elliott and Charlotte Denny, 'US Wrecks Cheap Drugs Deal', *The Guardian*, www.guardian co. uk./international/28/04/03
17. Quoted in *Biodiversity: What's at Stake*, CIIR Publications, Unit 3, Canonbury Yard, 190a, New North Road, London N1 7BJ, p. 24.

(now Syngentia), one of the world's largest agrochemical and biotechnology companies.[18] Already there are some seventy patents on the so-called 'golden rice'.

DOLLY

When the Roslin Institute in Scotland cloned the first mammal, Dolly, they applied for a broad spectrum patent which would give them exclusive rights over all cloned mammals. Almost immediately, they mounted a legal challenge against researchers at the University of Hawaii who were attempting to clone cows. Ian Wilmut and Keith Campbell claimed that the researchers in Hawaii used cloning techniques that are covered in their patent on Dolly.[19]

Dolly was hailed by many as a new wonder. Few commentators pointed out that it took 277 embryos to create her. Many of the pregnancies failed. Some of the lambs were stillborn or died at birth because they were unusually large. Then, in 2001, we found out that Dolly had arthritis at the relatively young age of five. Questions are being asked: Was it the cloning process that gave her a genetic defect? Dolly was put down in 2003.

DIAMOND VS CHAKRABARTY

Until recently, living organisms could not be patented. The decisive change in the push to patent living organisms began in the early 1970s. In 1971, Ananda Chakrabarty, a microbiologist working for General Electric, used genetic engineering techniques to develop a microbe that would help clean up oil spillage by devouring oil. Both the researcher and the company applied to the US Patent and Trademark Office (PTO) for a patent on the genetically engineered microbe. The

18. RAFI, 'Golden Rice and Trojan Trade Reps: A Case Study in Public Sector Mismanagement of Intellectual Property', RAFI *Communique*, Number 65, September/October 2000.
19. Philip Cohen, 'Who Owns the Clones?' *New Scientist*, 19 September 1998, p. 4.

Patent Office refused the application on the grounds that lifeforms could not be patented. Chakrabarty appealed to the Court of Customs and Patent Appeals (CCPA). To everyone's surprise, the CCPA in a three-to-two judgement reversed the PTO decision and granted a patent for the oil-consuming microbe, stating that 'the fact that the mico-organisms are ... alive ... [is] without legal significance.' [20]

The saga did not end there. The Patent Office challenged the CCPA's decision in the US Supreme Court. Before hearing the case, the Court advised the CCPA to examine a recent Supreme Court decision (*Parker vs Flook*) which stated that '[the courts] must proceed cautiously when we are asked to extend patent rights into areas wholly unforeseen by Congress.' [21] Despite this caution, the CCPA continued to uphold the patent. As a consequence, the Supreme Court was forced, in 1980, to address the issue of whether life could be patented.

Given the Court's stance in *Parker vs Flook*, most observers expected that the patent application would be refused. This did not happen. In June 1980 the US Supreme Court decided by a five-to-four majority that life was patentable. The ruling stated that the 'relevant distinction was not between living and inanimate things, but whether living products could be seen as "human-made inventions".' [22]

The Justices argued that the larger question, namely, whether life might be patented, should now be addressed by appropriate legislation in the US Congress. This never happened, so the *Chakrabarty* judgement opened the floodgates for patent applications on living beings. Andrew Kimbrell wrote: 'The complete failure by the Court to correctly assess the impact of the *Chakrabarty* decision may go down as among the biggest judicial miscalculations in the Court's long history.' [23]

20. Quoted in Kimbrell, op. cit., p. 193.
21. Ibid.
22. Andrew Kimbrell, op. cit., p. 195
23. Ibid.

One could not exaggerate the momentous nature of this decision. It constitutes a break with the way most cultures have viewed life down through the ages. The philosophical, ethical and legal bases on which the decision was reached are at variance with most of the cultural and religious traditions of the planet. Most cultures and ethical traditions make a clear distinction between living and inanimate realities. The Harvard biologist, Edward O. Wilson, would go much further in bonding humans with the rest of animate creation. In his book, *Biophilia*, he argues that during our evolutionary development we were hard-wired genetically to bond with other species in the living world. In the Prologue he uses a telling metaphor to illustrate the powerful attraction of other life forms: '[we] learn to distinguish life from the inanimate and move towards it like moths to the porch light'.[24] Nothing, and, certainly not the commercial demands of transnational corporations, should be allowed to blur or eliminate that vital distinction between life and non-life.

Furthermore, patents are derived from concepts of individual innovation and ownership which are foreign to many cultures where sharing of community resources and the free exchange of seeds and knowledge are promoted as crucial values. The concept of individual property rights to either resources or knowledge is alien to many indigenous people. In a patent-dominated world it is easy to forget that European and US agriculture was developed from plants and genetic resources freely imported from other countries. If justice means anything, Europe and America should repay their 'genetic debt' to the Third World.[25]

The simple fact is that Chakrabarty did not create 'his' bacterium. As Key Dismukes, a former director of the Committee on Vision of the National Academy of Science in the US

24. Edward O. Wilson, *Biophilia*, Harvard University Press, Cambridge, MA, 1984, prologue.
25. Jean-Pierre Berlan and Richard C. Lewontin, 'It's Business as Usual', *The Guardian*, February 22, 1999, p. 14.

observed, 'he merely intervened in the normal processes by which strains of bacteria exchange genetic information to produce a new strain with an altered metabolic pattern. "His" bacterium lives and reproduces itself under the forces that guide all cellular life.' [26]

Andrew Kimbrell believes that the US Supreme Court's decision has 'transformed the status of the biotic [life] community from a common heritage of the earth to the private preserve of researchers and industry.' He points out that the ruling has set the stage for increasing competition among multinationals as they vie for ownership and control of the planet's gene pool, patenting everything that lives, breathes, and moves. [27]

It is worth mentioning that this is not the first time that the judiciary had put the interests of corporations ahead of those of ordinary citizens. Fr Thomas Berry, a leading Catholic thinker on environmental issues, believes that 'from the beginning of the 19th century the legal profession and the judiciary in America bonded with the entrepreneurs and their commercial ventures, even at this early period, against the ordinary citizen, the workers and the farmers.' [28] He quotes from Morton Horwitz's book, *The Transformation of American Law 1780-1860*, to which reference is made in Chapter One. While we normally expect courts to uphold the law without favouring any particular group, Horowitz could write that 'by the middle of the 19th century the legal system had been reshaped to the advantage of men of commerce and industry at the expense of farmers, workers, consumers and other less powerful groups within society'.[29]

Patenting life certainly benefits corporations and not the general public, either in the First or Third World. Within a few

26. Quoted by Rifkin, op. cit. p. 46.
27. Kimbrell, op. cit. p. 200.
28. Thomas Berry, *The Great Work*, Bell Tower, New York, 1999, p. 143.
29. Morton J. Horowitz, *The Transformation of American Law 1780-1860*, Oxford University Press, New York, 1995, pp. 253-254.

short years, many genetically modified organisms, including viruses, plants and animals, have been patented in the US. The genes that are perceived to 'cause' many common illnesses have either been patented or had applications lodged for patents. Already Duke University in North Carolina has taken out a patent on the Alzheimer's gene, which they have licensed to Blaxo. The National Institute of Health has applied for a patent on the Parkinson's disease gene. Myriad Genetics, which is now owned by Novartis, has applied for a patent on a cardiovascular disease gene. Patent 5,633,161 on the melanoma gene is owned by Millennium Pharmaceuticals. Even a gene associated with obesity has now been patented by Millennium Pharmaceuticals and licensed to Hoffmann-LaRoche. These, and a host of other patents, will now be enforced in Europe since the Directive on the Legal Protection of Biotechnological Inventions was passed by the European Parliament on May 12, 1998. The EU Council of Ministers approved the Directive in the autumn of 1998.

Fortunately, the Dutch Government filed a nullity suit at the European Court of Justice against the Directive. Italy has also joined with the Dutch in opposition to the Directive. The Dutch challenge is based on a number of grounds, among them the fact that the Directive violates the basic rights of citizens by creating dependencies between patients and single companies (patent-holders).

The Italian Government recognised that patents on living organisms are morally objectionable to many people. Patenting promotes the view that life is a mere commodity. Most cultures and religions find this repugnant, especially when it includes human life.

Despite these challenges, the corporate world felt that it had a tough legal patent framework in place in the US and Europe. As a result, the number of applications for patents jumped astronomically from 150,000 per year in the late 1980s to 275,000 in 2001. In October 2000, there were patent applications on 126,672 human gene sequences. By February 2001, the figure had jumped a further 38 per cent to 175,624. The people who gain most from this are in the First World. For example, of the 26,000 patent applications filed to the African Intellectual Property Organisation, only 31 came from residents in Africa.[30]

LEARNING FROM THE BIBLE

The US Supreme Court's view of life also differs radically from the way life is understood, revered and cherished in the Judeo-Christian tradition. The first line of the Bible insists that everything was created by a living God: 'In the beginning God created the heavens and the earth' (Genesis 1:1). The text is very clear that all living beings, including human beings, are creatures of God.

Human beings have a special place in creation, as representatives or viceroys of God (Genesis 1:26). They show their dependence on God in the way they relate to God, to each other and to the earth. In the initial covenant between God and humanity (Genesis 1:28-31) humans were not allowed to eat flesh (Genesis 1:29). Even after the flood, when Noah was allowed to kill animals for food, there is a prohibition on consuming the animal's blood (Genesis 9:3-4). Blood, in the ancient Near-East, was considered to be the seat of life. The Old Testament scholar Gerhard von Rad writes: 'even when man slaughters and kills, he is to know that he is touching something which, because it is life, is in a special manner God's property'.[31]

The first account of creation goes on to teach that all beings have their own inherent value. This dignity derives from the fact that they are created by God (Genesis 1:12.19-25). This inherent dignity of creatures increases and intensifies the higher one moves up the chain of being. In Genesis 1:21-22

30. Madeleine Bunting, 'Profits That Kill', *The Guardian*, February 12, 2001, p. 19.
31. Gerhard von Rad, *Genesis*, SCM Press Ltd, London, 1961, p. 128.

God blesses creatures that live in water, and the birds.

In the second account of creation the 'man' is given the privilege of naming the animals (Genesis 2:19-20). The text recognises that all creatures, including humans, have a common origin. We are all created from the soil and are all part of the web of life. God invites the 'man' to name the animals and thereby incorporate these creatures into the human environment. While this gives humans dominion over other creatures, it does not give the right to oppress and exploit. Rather, such dominion is to be patterned on God's own care and sovereignty, as expressed in Psalm 72:4-6 where the righteous king combines concern for the poor with care for the creatures of the earth.

Furthermore, in the Judeo-Christian tradition, creation is an all-encompassing activity. It is not a once-off action in the distant past by a mechanistic God who immediately abandons the world to its own devices. From the time of Origen (*ca* 185– *ca* 254, A.D.) creation was understood as a continuing reality. Catholic theology affirms that God's initiative in creation is not confined to the initial moment of creating the universe out of nothing. Catholic doctrine has always stated that God is constantly involved with creation. God is perceived as living in each creature, as holding together the web of life and as leading all creation into the future (Psalm 104). In the Catholic theological tradition, 'creation is not an artifact. It is a gift, not of improved or altered being, but of being, pure and simple.' [32]

The Bible does not share the reductionist myopia of the US Supreme Court that sees life as an isolated entity and as a product of human industry. In the tradition of Thomas Aquinas, all being is indebted to God for its existence and continuation in existence. Underlying all action in the world and human affairs is the God who keeps us in being and

---

32. Denis Carroll, 'Creation', in Dermot Lane et al. (eds.): *The New Dictionary of Theology*, Gill and Macmillan, 1987, p. 250.

enables our action.[33] Modern theologian Jürgen Moltmann writes: 'if we want to understand what is real as real, and what is living as living, we have to know it in its own primal and individual community, in its relationships, interconnections and surroundings'.[34]

Patenting is a fundamental attack on this understanding of life as interconnected, mutually dependent and a gift of God given to all. It opts instead for an atomised, isolated understanding of life. It is also at variance with the Judeo-Christian conviction that freedom, openness and possibility are the hallmarks of life in God's creation.

The Bible also recognises that humans are companions and stewards of other creations in the community of life (Genesis 2:15). In Genesis 2:15-17, God settles the 'man' in the Garden and invites him to cultivate it and care for it. The text goes on to place certain limits on the man's use of the natural world. Yahweh God gave the man this admonition: 'You may eat indeed of all the trees in the garden. Nevertheless, of the tree of the knowledge of good and evil you are not to eat, for on the day you eat of it you shall surely die.' (Genesis 2:16-17)

But stewardship does not mean that humans are inventors or owners of life, or that they can dominate and exploit everything in creation. In fact, it challenges and repudiates that view. God, and only God, is the Creator of life, and all life, including human life, and every creature is dependent on God. The Bible is very critical of those who, puffed up with arrogance, refuse to recognise that they are creatures and, thus, dependent on God. In the story of the Tower of Babel (Genesis 11) humans repudiate God's sovereignty and attempt to storm heaven under their own steam. I think it would not be misrepresenting the meaning of this text to interpret any claim to own life as usurping the divine prerogative as

33. Ibid., p. 251.
34. Jürgen Moltmann, *God in Creation*, Harper and Row, San Francisco, 1985, p. 3.

author of life.

Living organisms are not merely 'gene machines' to be manipulated and exploited for profit. This is why, after the US Patent and Trademark Office patented the first animal in 1987, a group of twenty-four religious leaders issued the following statement:

> The decision of the US Patent Office to allow the patenting of genetically engineered animals presents fundamental dangers to humanity's relationship with the natural world. Reverence for all life created by God may be eroded by subtle economic pressures to view animal life as if it were an industrial product invented and manufactured by humans. [35]

In his encyclical on social justice, *Sollicitudo Rei Socialis*, Pope John Paul II interprets the Genesis 2:16-17 text as placing limitations on humans' use of the natural world:

> The dominion granted to man by the Creator is not an absolute power, nor can one speak of a freedom to 'use and abuse', or to dispose of things as one pleases. The limitations imposed from the beginning by the Creator himself and expressed symbolically by the prohibition not to 'eat of the fruit of the tree' shows clearly enough that, when it comes to the natural world, we are subject not only to biological laws, but also to moral ones, which cannot be violated with impunity.[36]

Surely, the limitations referred to by the Pope must include respect for the genetic integrity of other species, and must preclude any claim to ownership over life.

The Pope again raised the question of genetic engineering in his World Day of Peace message for 1999. He stated: 'recent developments in the field of genetic engineering present a

---

35. Quoted in Kimbrell, op. cit., p. 201.
36. *Sollicitudo Rei Socialis*, no. 34.

profoundly disquieting challenge. In order that scientific research in this area may be at the service of the person, it must be accompanied at every stage by careful ethical reflection, which will bring adequate legal norms safeguarding the integrity of human life. Life can never be downgraded to the level of a thing.' But this is exactly what patenting does; it denies the fundamental notion that life is primarily a gift and treats it like an inanimate object.

The Pope returned to the issue in an address he gave to the members of the 'Jubilee 2000' Debt Campaign. In the midst of a talk on Third World Debt he had this to say:

> The Catholic Church looks at the situation with great concern, not because she has any concrete technical model of development to offer, but because she has a moral vision of what the good of individuals and the human family demand. She has consistently taught that there is a 'social mortgage' on all private property, a concept which today must also be applied to 'intellectual property' and to 'knowledge'. The law of profit alone cannot be applied to that which is essential to the fight against hunger, disease and poverty.

The Church of Scotland has also come out against patenting living organisms. Its opposition is grounded on theological and ethical arguments. A Report commissioned by the Church stated:

> Living organisms themselves should therefore not be patentable, whether genetically modified or not. It is wrong in principle. An animal, plant or microorganism owes its creation ultimately to God, not human endeavour. It cannot be interpreted as an invention or a process, in the normal sense of either word. It has a life of its own, which inanimate matter does not. In genetic engineering, moreover, only a tiny fraction of the makeup of the organism can be said to be a product of the scientist. The organism is still

essentially a living entity, not an invention. A genetically modified organism is in a completely different category from a mouse trap.[37]

OPPOSITION TO PATENTING LIVING ORGANISMS

Opposition to patenting living organisms has come from many quarters, including tribal and peasant people, scientists and religious people. The arguments against are based on economic, social, scientific and ethical considerations. The Union of Concerned Scientists in the United States, for example, has consistently opposed the patenting of living organisms. They argue that patents make important products more expensive and less accessible.

Sir John Sulston, the British scientist who won the Nobel Prize for medicine in 2002, is opposed to patenting life. Sir John exemplifies all that is best in British scientific tradition. He worked in a university where he was able to devote thirty years to studying a hermaphrodite nematode, without having to seek corporate funding for his research and consequently being diverted to respond to the corporation's agenda rather than pursuing scientific discoveries. His painstaking research led him to find out how cells develop and die under instruction from their genes. Medical scientists studying how cancers develop need this kind of accurate information. This explains why an expert on nematodes shared the 2002 Nobel Prize in medicine.

In collaboration with Bob Waterston in the US, Sulston promoted the publicly funded and publicly accessible codification and sequencing of the human genome. Writing in the *Guardian,* Andrew Brown, author of *In the Beginning Was the Worm,* comments:

37. Church of Scotland, 'Supplementary Reports to the General Assembly and Deliverance of the General Assembly 1997', May 1997. Quoted in *Biopatenting and the Threat to Food Security, A Christian and Development Perspective,* CICSE, International Cooperation for Development and Solidarity, p. 17.

Sulston believes passionately that the information on the genome sequence must be freely available and that it is wrong to patent human gene sequences, both morally and scientifically. It is morally wrong because human genes are discovered, and not invented, while the patent on a discovery blocks all invention in that area. If you patent a discovery which is unique, say, a human gene or even just one particular function of a human gene, then you are actually creating a monopoly and that's not the purpose of the world of patent. Indeed, the purpose [of patents] is to cause inventors to compete with each to get better products. So mousetraps are in one category, human genes are in another! says Sulston. [38]

It was also clear to Sulston that, in order to achieve results in his work, he depended on the collaboration of other scientists. He realised that he could not have made significant breakthroughs in his field without building on the work of other scientists. His studies of the worm's cell lineage would not have been possible without the very detailed physical map of the worm produced by other researchers. Brown insists that '[t]here is no doubt that Sulston believes that DNA patents are immoral. But he is just as keen to argue that they damage science.' [39] Finally, Sulston has not become an extraordinarily rich man like many other researchers in molecular biology and genetics. He believes in working for the common good, the betterment of humankind and the increase in knowledge that should be available to everyone.

MANY THIRD WORLD PEOPLE OPPOSE PATENTING

Patenting life is not seen favourably in the Third World. Isidro Acosta, the president of the Guaymi General Congress in Panama, was shocked and outraged when he heard that the

---

38. Andrew Brown, 'One Man and his Worm', *The Guardian*, October 9, 2002 http://www.guardian.co.uk/g2/story/0,3604,807110,00.html
39. Ibid., p. 4.

US government was attempting to take out a patent on a virus taken from the cell line of a twenty-six-year old Guaymi woman in Panama. Acosta stated: 'It's fundamentally immoral, contrary to the Guaymi view of nature ... and our place in it. To patent human material ... to take human DNA and patent its products ... violates the integrity of life itself, and our deepest sense of morality'.[40]

Peasant farmers are also opposed to patenting. At a meeting of a network of peasant organisations called MASIPAG (Magsasaka at Siyentipiko Para sa Ikauunlad ng Agham Pang-Agrikultura) on the island of Negros in the Philippines in January 1999, seven thousand people protested against the patenting of life. They denounced the Intellectual Properties Treaty of the WTO. In the following years MASIPAG produced pamphlets in English and various Filipino languages opposing genetic engineering and patenting.[41]

A similar meeting of fifty peasant, indigenous and environmental organisations took place in Quito, Equador, in January 1999 to review contemporary developments in agricultural biotechnology. As a result of their deliberations they published the Latin American Declaration on Transgenic Organisms. This document rejects genetic engineering and patenting. It states: 'genetic engineering is a technology driven by commercial interest. It is not necessary. It forces us to become dependent on TNCs which control it, putting our autonomy to take decisions about production systems and food security in real danger. In the field of agriculture, especially, there are traditional and alternative technologies which do not pose such risks and which are compatible with the conservation of biodiversity'.[42]

The South Asian Network on Food, Ecology and Culture (SANFEC) organised a workshop on patenting in Tangil,

---

40. Jeremy Rifkin, op. cit., p. 59.
41. MASIPAG 3346 Aquila, Rhoda Subdivision, Los Banos, 4030 Laguna, Philippines. E-mail: zcom.com masipag@mozcom.com
42. Quoted in Kimbrell, op. cit., p. 200.

Bangladesh. The final statement on intellectual properties rejects patenting:

> South Asian communities are historically premised on the deep sense of moral, religious and cultural values. The region is inhabited by multi-ethnic, multi-religious and large indigenous communities. All trees, crops, animals, birds, organisms, and soils are an inalienable part of our worship, our rituals, our celebrations, our joys, our culture of sharing and our loving affinity to each other. Our region is replete with hundreds of thousands of sacred groves where trees and plants are worshipped by people. We have a long history of spiritual and political movements where Sufis, Saints and various bhakti traditions have fought to preserve the integrity of Nature in her multiple expressions, including the beauty of life forms.
>
> Such gifts must be cared for and respected and only then do we gain our moral rights to use them for our livelihood needs. The human as omnipotent consumer, that owns, controls, mutates, displaces and destroys the environment, through privatizations, colonizations and now through intellectual property rights (IPRs) in life-forms, is totally against our culture. We are strongly opposed to non-recognition of the rights of other cultures to live on their own historical premises and principles.

Some farmers in First World countries also are opposed to patenting. In Canada and the US, Monsanto engaged the services of an investigative agency to gather information on over 1,000 farmers they consider are cheating on patented seeds.[43] The affected farmers have coined a new word – 'bioserfs' – to capture the feudal relationship which now exists between many seed companies and farmers. The experience of the Canadian canola farmer, Percy Schmeiser, mentioned

---

43. Cathryn Atkinson, 'Seeds of Doubt', *The Guardian*, February 2, 2000, p. 4

in Chapter Five, ought to alert other farmers to what can happen when seeds are patented. Monsanto filed a lawsuit for patent infringement because some genetically engineered canola was found on his land. Schmeiser is adamant that he did not plant Monsanto's GE canola. He insists that he is the aggrieved party because his non-GE seeds, which he had been developing for fifty-three years, were contaminated with Monsanto's GE canola from a neighbouring farm where GE canola seeds were used. In 2000 the court ruled against Schmeiser. It did not matter how the GE seeds arrived on the farm, whether by cross-pollination or whether by being blown in on the wind. The very fact that the plants were on his property meant that he was guilty. The judge ruled that all the profits from his 1998 harvest must go to Monsanto, even from the fields where no GE seeds were found.

Schmeiser believes that one of his neighbours used a confidential hotline to alert Monsanto officials to the fact that GE seeds were sprouting on Schmeiser's farm. By 2002, Schmeiser had spent $125,000 in lawyers' fees and an appeal will cost him another $50,000. He believes that Monsanto is intent on gaining complete control of the staple crops of the world by controlling seeds. As we saw in Chapter Five, in the past decade Monsanto has spent millions of dollars buying up seed companies all round the world. Although patents have run out on Monsanto's flagship chemical Roundup Ready, farmers who used Monsanto's GE crops will be forced to use Roundup Ready.[44] In the light of Schmeiser's experience, it is clear to many people across the world that patenting seeds and animals is now seen as a major economic, development and ethical issue.

PATENTING WILL HINDER PROGRESS IN SCIENCE AND MEDICINE

Opponents of patenting also believe that a patenting cul-

44. Interview with Percy Schmeiser, 'Seeds of Discontent', *WorldWatch*, January/February 2002, pp. 8-10.

ture will promote a climate of secrecy in science and hinder the normal exchange of information that is essential in order to promote scientific research. Scientific information and the materials required for research will become more expensive and difficult to obtain if one corporation owns a patent on it. In practice, this will deter rather than promote research.

With the passing of the Biopatenting Directive in the European Parliament in May 1998, a patent owner can now decide who will be allowed to use the gene or gene sequence for developing a diagnosis, therapy, medicine or transgenic organism. Therefore, it is obvious that patenting will actually hinder research. For example, in 1997 a British and US team of researchers were working together on isolating and decoding the gene for breast cancer. Once the gene was isolated the US team patented it and effectively pushed their British colleagues out of the race because the royalties for the patent were too high.[45]

A research culture focused on patenting will also mean that scientific research will no longer be undertaken simply to increase our understanding of the world, to search for truth or to promote the public interest. Even today it appears that scientific research in genetics is driven more by the search for corporate profit and patent control than by concern for human or planetary well-being. Many companies are applying for patents to scare off competition by 'staking out an area of research'.[46]

Sheldon Krimsky of Tufts University in Medford, Massachusetts, examined 789 biomedical papers published by 1,105 scientists based in Massachusetts Universities in 1992. In 34 per cent of the papers, at least one of the authors stood to gain financially from the results they were publishing, either because they held a patent, or were employed by a biotech

---

45. Greenpeace paper prompted by the EU Parliament's decision on the Directive on the Protection of Biotechnological Inventions, July 1997.
46. Madeleine Bunting, 'The Profits That Kill', *The Guardian*, February 12, 2001, p. 16.

company that was exploiting the research. An even greater cause for concern is the fact that none of the 267 papers where the author stood to gain financially from the research, mentioned that fact. Krimsky discovered the financial links only by trawling through databases of US patents and registers of corporate officers for the names of the authors of the 1,105 papers.[47]

In January 2000, James Meek, a columnist in *The Guardian*, reported that an American company which has 'patented' two human genes for breast cancer screening is threatening the work of fifteen publicly funded British laboratories that perform a genetic test for half the price the American company charges.[48] So great is the perceived threat to medical research that a group of American doctors and scientists has issued a protest saying: 'the use of patents or exorbitant licensing fees to prevent physicians and clinical laboratories from performing genetic tests limits access to medical care, jeopardises the quality of medical care and unreasonably raises its costs.' [49]

In September 2001, thirteen of the world's leading medical journals, including the *Lancet*, the *New England Journal of Medicine* and the *Journal of the American Medical Association*, mounted a concerted attack on pharmaceutical companies, accusing them of 'distorting the results of scientific research for the sake of profits'.[50] They claimed that drug companies 'tie up academic researchers with legal contracts so that they are unable to report freely and fairly on the results of the drug trials.' [51] The *British Medical Journal* returned to this theme in May 2003. The editor chided the drug companies and called

47. Vincent Kiernan, 'Truth Is no longer Its Own Reward', *New Scientist*, March 1, 1997.
48. James Meek, 'US Firm May Double the Cost of UK Cancer Checks', *The Guardian*, January 17, 2002.
49. Julian Borger, 'Rush to Patent Genes Stalls Cures for Disease', *The Guardian*, December 15, 1999, p. 1.
50. Sarah Boseley, 'Drug Firms Accused of Distorting Research', *The Guardian*, September 10, 2001, p. 2.
51. Ibid.

on doctors and the industry to try to forge 'less grubby' ties.[52]

Connections between doctors and drug companies have very worrying implications for public health. They should be investigated immediately by competent and well-resourced government agencies and the medical profession itself. The chance of this happening is close to zero. In today's world TNCs are monarchs who are regularly wooed by governments and dispense largesse to many doctors in the form of free trips to international drug company-sponsored conferences. Though the call by the *British Medical Journal* for relations between doctors and drug companies to become less grubby was courageous and timely, sadly, it received little coverage. Pressure by corporations on researchers will further deepen the distrust that many feel about the reliability of in-company research trials, where billions of dollars may either be made or lost depending on whether a drug proves successful or has to be discarded.

PRIVATE RESEARCH IS MARGINAL TO BREAKTHROUGHS IN AGRICULTURAL AND MEDICAL RESEARCH

The biotech industry would like the public to believe that they have funded the bulk of medical and agricultural research and are, therefore, entitled to charge patenting royalties. The reality is, in fact, very different. Jean-Pierre Berlan, Director of Research at the National Agronomic Research Institute (INRA), and Richard C. Lewontin, holder of the Alexander Agassiz chair of zoology and professor of population genetics in Harvard, refuted these claims.[53] They stated that we owe the unprecedented increase in yields in the industrial countries, as well as the Third World, to the free movement of knowledge and genetic resources. (Yields have increased four- and five-fold in two generations, after taking twelve to fifteen generations to double and being, no doubt,

52. James Meike, 'Medical Journal Turns on Drug Firms, *The Guardian*, May 20, 2003, p. 14.
53. Richard C. Lewontin, *The Guardian*, February 22, 1999.

unchanged for thousands of years before that.) The contribution of private research has been marginal, in the US as elsewhere.

The US has developed a hybrid maize, but, in the course of the 1970s, nearly all the hybrids in the US corn belt were the result of crossing two public strains – from the universities of Iowa and Missouri. Public research and public research alone, does all the basic work on improving the population of plants on which everything depends. Research work is being hampered by the privatisation of knowledge, genetic resources and the techniques for their use. Tired of paying royalties on genetic resources that were snatched from them in the first place, many countries in the Southern hemisphere are now trying to stop their circulation.

In the neo-liberal fervour of the Thatcher and Reagan years, the pressure to privatise public knowledge has gathered momentum. Within a few short years, the private sector has taken over public research. For example, less than 6 per cent of all public sector patents were surrendered via exclusive license to private companies in 1981. By 1990 the figure had jumped to 40 percent, and it was estimated that all the intellectual property accruing to US universities and government agencies would be controlled by TNCs on an exclusive access basis by the end of the twentieth century.[54]

MOST RESEARCH IS FUNDED BY TAXPAYERS AND CHARITIES

In response to the 'No Patents, No Cures' argument, it is important to point out that much of the improvement in biomedical knowledge and procedures has been funded by taxpayers and charitable organisations. Vast amounts of public funds have been allocated to cystic fibrosis and breast cancer research. It will be ironic if public medical institutions have to pay royalties to biotech companies in order to use screening tests that were developed using knowledge that was

54. Mooney, op. cit. p. 140.

gained in these institutions.

PATENTING WILL INCREASE THE COST OF HEALTH CARE

The group, Disabled Against Animal Research and Exploitation (DAARE) believes that the European Patent Directive will increase health costs and place the discoveries of publicly funded research in the hands of private corporations. In 1997, the Central Manchester Regional Genetics Centre received a bill from a Toronto-based biotech company demanding a $5,000 licence fee and a further $4 royalty each time the Centre uses a cystic fibrosis gene screening test on which the Canadian company has filed a patent application. Before the European Parliament voted on the Biopatenting Directive in May 1998, the Centre paid no royalty since the patent operated only in Canada. Now, existing patents will operate within the EU. Since the Centre cannot afford such costs patients will suffer.

The patent application of the US Biotech company, Myriad, will touch the lives of a larger segment of the population, especially women. Myriad has applied for a patent on the breast cancer gene BRCA-1, as well as on all therapeutic and diagnostic applications that result from knowledge of the gene. If this patent is granted, the company will be allowed to charge patients every time a diagnostic screening is performed. At present, it costs the National Health Service in Britain £600 to screen patients for two breast cancer genes BRCA-1 and -2, and £30-35 for each subsequent test. Myriad Genetics, on the other hand, charges £1,500 to screen for the gene, and £300 for subsequent tests.

Such costs would be prohibitive and would restrict access to these tests to the super-rich. Staff at the Regional Genetics Service of Central Manchester Healthcare wrote to all the members of the European Parliament in July 1997, pointing out that patenting genes would make 'the possibility of genetic testing for disorders such as heart disease or breast cancer so prohibitively expensive it would be beyond the scope of the NHS [National Health Service]'.[55]

In 1998, a US company applied to patent the bacterium that causes meningitis. If a new vaccine is found to deal with the disease, patients would have to pay a royalty every time the vaccine was administered.[56]

The World Medical Association (WMA) declared patenting life forms unethical in 1998. This organisation, representing doctors and scientists in seventy-seven countries, is opposed to patenting life-forms because such patenting is aimed at maximising profits rather than making treatment available to patients. The association believes that doctors have an ethical obligation to share their skills and knowledge with their colleagues. Patenting will undermine these obligations because a patenting regime would limit the dissemination of knowledge, especially if scientific results were delayed in the hope of obtaining a patent.[57]

PATENTS WILL PROMOTE UNSUSTAINABLE AGRICULTURE

Patents promote unsustainable and inequitable agricultural policies. A disastrous decline in genetic diversity could result from patenting crop species. The development of genetically uniform organisms would make it easier for corporations to maintain their patent claims. Biotech companies holding broad spectrum patents on food crops will encourage farmers to grow modified varieties with promises of greater yields and disease resistance. However, numerous examples worldwide show the 'improved' crops have failed to hold up to corporate promises, and have led to the loss of the rich diversity of traditional crop varieties.

55. *No Patents on Life! A Briefing on the Proposed EU Directive on the Legal Protection of Biotechnological Inventions,* September 1997, The Corner House, P.O. Box 3137, Sturminster, Newton, Dorset, DT10 1YJ, p. 7.
56. David King and Paul Brown, 'Firm Attempts to Patent Meningitis Bacteria', *The Guardian,* May 7, 1998, p. 1.
57. Paul Brown, 'Doctors Urged to Shun Patents on Life-forms', *The Guardian,* May 8, 1998.

The patenting of seeds will give enormous economic power to a small number of agribusiness corporations selling their wares on the global market, and not selling them cheaply. The insect-resistant maize hybrid produced by Pioneer Hi-Breed requires access to 38 different patents controlled by 16 different patent holders.[58] In 1992 a biotech company called Agracetus, which is a subsidiary of the US corporation W. R. Grace, was given a patent for all genetically engineered cotton plants. Thus one company has a monopoly on new cotton seed strains.[59]

In addition, farmers will be forced to pay royalties on succeeding generations of plants and animals that they buy or produce. It will be illegal to save seeds from the previous harvest without permission and payment. This will make farmers totally dependent on transnational agribusiness corporations.

The impact on Third World countries will be devastating, with a further flow of financial resources from the South to the North. Patenting will institutionalise the dependence of Southern agriculture on Northern agribusiness corporations in whose hands the flow of scientific information and new agricultural technologies will be concentrated. As a result, instead of feeding the hungry in the South as the agribusiness corporations claim, the new situation could create food shortages and famine.

BIOPIRACY

The patenting of Third World genetic resources by First World corporations or institutions represents theft of community resources. Much of the raw material used in genetically engineered food and medicinal plants is found in Third World countries. In recent years, biotechnology companies have been collecting this material, patenting their products

58. Greg Horstmeier, *Farm Journal*, October 1996. Quoted in Mooney, op. cit. p. 139.
59. Celia Deane-Drummond, op. cit., p. 87.

and in the process making huge profits. Even before the advent of biotechnology, Eli Lily was in a position to make millions of dollars by developing a drug to treat some cancers from a plant called the rosy periwinkle which is found in the rainforest of Madagascar. In 1993 alone, the company made $160 million profit in sales but did not contribute one dollar to Madagascar where the plant was found.

Patenting will intensify and exacerbate the plunder of the Third World's natural resources. Microorganisms, plants, animals and even the genes of indigenous people have been patented for the production of pharmaceuticals and other products. It is nothing short of robbery to design international mechanisms that force developing nations to pay royalties to the wealthy industrial nations for products derived from their own natural resources.

Most of the world's germ plasm for crops and animals is held in seed banks either in the North or controlled by the North, though it originated in the South – germ plasm is the part of the seed that contains the hereditary material. To appropriate this, through patenting or Plant Variety Protection (PVP) legislation, constitutes a new form of colonialism. This time it is not merely the gold, silver or labour of people that is being colonised, but life itself. Biotech scouts have used the knowledge of indigenous plants which local people have accumulated over centuries, in their search for plants and animals which may have an agricultural or medical use, and then patented these products.

The immorality of such behaviour is magnified even further when one remembers that the species and genetic diversity available today occurs because countless generations of Third World farmers protected, preserved, propagated and shared these species freely with others. Vandana Shiva, an Indian scientist and activist, points out that

> the common pool of knowledge has contributed immeasurably to the vast agricultural and medicinal plant diversity

that exists today. Thus, the concept of individual property rights to resources or to the knowledge, remains alien to the local community. This undoubtedly exacerbates the usurpation of the knowledge of indigenous people with serious consequences for them and for biodiversity conservation.[60]

Now, all this richness is destined to be privatised for the exclusive benefit of Northern corporations. This will give them huge control over the food supply of our world. At present ten corporations control 32 percent of the commercial seed market, valued at $23 billion, and 100 percent of the market for genetically engineered seeds.[61]

The Neem Tree

Two examples drawn from India and West Africa illustrate what is happening now. The *neem* tree is found all over India. Farmers and traditional healers have used this tree for a variety of purposes for hundreds of years. In ancient Sanskrit texts the tree is called *sarva roga nivarini* (the curer of all ailments), while Indian Muslims refer to it as *shajar-e-mubarak* (the blessed tree). The fact that everyone, even the poorest people, had access to its beneficial properties is captured by the Latin name, *Azadirachta indica* which is derived from the Persian name meaning 'free tree'.

However, it is possible that Indian citizens will soon be required to pay royalties on the products produced from the *neem*, since a patent has been granted to the US company W.R. Grace, on a compound in the tree (azadirachtin) for the production of a bio-pesticide. In 1993, over five hundred thousand South Indian farmers rallied to protest against foreign patents on plants such as the *neem*, and launched a nationwide resistance movement. Transnational corporations can make huge profits on their 'discoveries', while depriving

60. Vandana Shiva 'The Enclosure of the Commons', *Third World Resurgence*, August 1997, p. 6.
61. Vandana Shiva *Stolen Harvest*, Zed Books, London, 2000, p. 9.

the communities which have fostered this knowledge for centuries of the beneficial properties of their own flora and fauna.

In West Africa the berry *brazzein* (pentadiplandra brazzeana) is renowned for its sweetness. This berry is much sweeter than sugar and, unlike other non-sugar sweeteners, it does not lose its taste when it is heated. This makes it an ideal candidate for the sugar-free food industry which is worth about $100 billion a year. A US researcher from the University of Wisconsin who saw people and animals eating the berry applied for a US and European patent on the protein isolated from the berry. The drive to create a genetically engineered organism to produce *brazzein* is under way. This will eliminate the need to grow the berry in West Africa. Naturally, given the market for such a sweetener, there is huge commercial interest in the project.

Most fair-minded people would consider it totally bizarre for the University of Wisconsin to claim that *brazzein* is 'an invention of a UW-Madison researcher'. There are no plans to share any of the benefits of the discovery with the people of West Africa who nurtured this plant for generations.[62] The knowledge, innovation and efforts of these communities is neither acknowledged nor rewarded. Such biopiracy on the part of Northern institutions and corporations should not be legitimised by cleverly worded patent legislation.

Because of its location in the tropics, the Philippines is a country very rich in flora and fauna. Before the destruction of the Philippine forests, they were home to almost 13,500 plant species or almost 5 percent of all the plant species in the world.[63] 558 bird species have been found in the Philippines and of those, 171 are found nowhere else in the world. In marine ecosystems, 4,951 species of marine plants and ani-

62. Grain: 'Patenting, Piracy and Perverted Promises: Patenting Life, the Last Assault on the Commons', Girona 25, pral, F-08010 Barcelona, Spain, pp. 5-6.
63. Department of Environment and Natural Resources, *Philippines Biodiversity – an Assessment and Action Plan*, The Philippines and United Nations Development Plan, Manila, 1977, p. 36.

mals have been found.[64] A further 1,616 species of flora and 3,675 species of fauna are found in Philippines lakes, rivers, marshes and swamps.[65] The race to 'discover' and patent many of these is already underway. The Philippine sea snail, *conus magus*, produces one of the world's most powerful painkillers. This snail has now been patented by the US transnational corporation Neurex, Inc.

Even when a corporation enters into a deal with a country such as the well-publicised arrangement between the chemical company Merck & Co and Costa Rica, the benefits which the host nation receives are paltry. Merck has agreed to pay $1 million to the National Biodiversity Institute in Costa Rica in return for being allowed to collect micro-organisms, plants, insects and animals in one of the areas of greatest biodiversity on the planet. Over the long-term, the contract could mean billions of dollars profit for Merck. All they will have to do is pay a pittance of $1 million to a Costa Rican institute. It is worth noting that the indigenous people who live in the forest and whose knowledge of the plants and animals will be crucial in making the agreement work, are not included in the deal.

TRIPs came under sustained attack at the WTO Seattle meeting in November 1999. The US trade representative Charlene Barshefsky and the director-general Mike Moore from New Zealand tried to get a statement from the meeting. The vast majority of Third World countries were excluded from the decision-making meetings. The African countries were so exasperated at the way they were being treated that they issued a statement that the whole meeting lacked transparency and that they were excluded from discussing issues which were vital for their future. Little wonder that the meeting ended in a shambles. The proposed review of the implementation of TRIPs never happened.

Organisations like the WTO are not easily thwarted. By

64. Ibid. p. 55.
65. Ibid.

Spring of 2001 the WTO personnel had regrouped and were laying the groundwork for another round of trade negotiations to further liberalise trade. This included agricultural produce, despite the caution from many knowledgeable commentators who have linked the increase in infectious diseases, like foot-and-mouth among ruminants, with the increased unregulated movement of plants and animals over great distances. William Cashman, an Irish veterinary surgeon, claimed that 'the active promotion of "free-trade" has facilitated the movement of animal diseases, given the distance that modern transport can move animals and products over a short time. Many EU inspection missions within Europe have expressed dissatisfaction with transit monitoring measures to protect animal and human health, but to date no effective action has been taken to strengthen procedures'.[66]

THE NEED TO RE-WRITE TRIPs

Many Third World countries and NGOs had hoped that the WTO Cancún meeting would review the article on TRIPs and re-write it in a way that protects human health and vulnerable subsistence farmers in the Third World, and also protects the environment. The main plank of such a review would be the affirmation that all living beings ought to be considered the common property of our humanity and our earth. The hypocrisy of the current TRIPs position is that it fails to protect the genetic resources of the South while at the same time it facilitates the patenting of genetic resources which will benefit Northern multinational corporations. This is not free-trade; rather, it is a ploy to create global monopolies.

It is true that in its present form TRIPs Article 27.3(b) allows members to exclude from patentability plants and animals other than micro-organisms, and biological proc-

---

66. William Cashman, MVB, MRCVS: 'Free Trade and Disease', letter to *The Irish Times*, March 2, 2001.

esses essential for the production of plants and animals other than non-biological and micro-biological processes. The trouble is that TRIPs 27.3(b) demands that member states enact legislation which is tantamount to patenting. It states that member states shall provide for the protection of plant varieties either by patents or by an effective *sui generis* system or by any combination thereof.

Despite the fact that the patenting of organisms is excluded from the GATT agreement, the whole tone of the document supports patenting. In the past few years the US government has put pressure on many Third World governments to adopt *sui generis* legislation along the lines set down by the Geneva-based Union for the Protection of New Varieties of Plants (UPOV). This UPOV system operates under a convention set up in 1978 and amended in 1991. According to Julian Oram, a researcher at the International Famine Centre, University College Cork, the UPOV guidelines treat the South's biodiversity as a part of the 'heritage of mankind' and therefore it is freely available for scientific and commercial use. Once a corporation has acquired this material and 'transformed' it through genetic engineering techniques, they can claim property rights on the basis that they have made an 'invention'. Oram writes: 'having done this the "free heritage of mankind" plundered from the fields and forests of local communities could be sold back to them as a commodity'.[67] Obviously, this approach helps breeders but not the farmers and, not surprisingly, the corporations are promoting it.

Plant Variety Protection legislation (PVP), while not as strong as patenting, protects the genetic makeup of a specific plant variety. The criteria for protection are also somewhat different. They include novelty, distinctiveness, uniformity, and stability. PVP laws can also provide exemptions for breeders, allowing them to use protected varieties for further

---

67. Julian Oram, 'The TRIPs Agreement and Its Implications for Food Security', unpublished talk, Dublin, September, 1999.

breeding. In certain highly restricted circumstances, farmers can save seeds from their harvest. However, in many Third World countries like the Philippines, the PVP legislation (Senate Bill no. 1865, co-authored by Senators Manuel Villar, Juan Flavier and Sergio Osmena III) extends the rights of breeders to the farmers' harvest and the direct product of that harvest. If, for example, a farmer sowed a field with a protected variety without having paid a royalty, the company that produced the seed had a right to claim ownership of the harvest.[68] Many environmentalists feel that once enact PVP legislation is enacted, agribusiness corporations already have their foot in the door on the way to a full-scale patent regime.

It is crucial that Third World countries and NGOs from the North and South call for a root-and-branch review of Article 27.3(b). It is important to oppose patenting, PVP and Material Transfer Agreement (MTAs) in order to protect the biological resources of the South from predatory Northern TNCs who are attempting to gain monopolies in the seeds of many staple crops.[69]

The Article should be amended as follows: member countries shall exclude from patentability all life forms, including plants, animals, micro-organisms and parts thereof; and also exclude from patentability all natural processes for the production of plants, animals, micro-organisms and all living beings. There must also be a concerted effort to rescind the one-size-fits–all approaches to patenting which is vigorously promoted by the multinationals.

68. Robert Ali Brac De La Perriere and Franck Seuret, *Brave New Seeds: The Threat of GM Crops to Farmers*, Zed Books, London, 2000, p. 99.
69. MTAs are legal contracts between two or more parties specifying the conditions under which material like seeds can be exchanged. According to John Barton and the late Wolfgang Siebeck, 'failure to perform what is promised in an MTA is a breach of contract which gives one party the right to bring action against the other party, such as suing for damages'. Barton, J. H. and Siebeck, W.E.: 'Material Transfer Agreements in Genetic Resources Exchanges: the Case of the International Agricultural Research Centers', *Issues in Genetic Resources*, No 1, International Plant Genetic Resources Institutes, Rome, May 1994.

THE CONVENTION ON BIOLOGICAL DIVERSITY (CBD)

In 1989, the United Nations Environment Programme (UNEP) set up a working group to design international laws and conventions to protect biodiversity. This was in response to the current extinction spasm which is having such a devastating effect on all life on the planet. It is estimated that up to 40,000 species are pushed across the abyss of extinction each year. At the Earth Summit in 1991, 150 countries signed the Convention on Biological Diversity (CBD). By the year 2000, 170 nations had signed up. Ireland signed in 1996.

The objective of the CBD is to protect biodiversity and to ensure that there is a fair and equitable distribution of financial benefits derived from these biological and genetic resources. For this reason, CBD is more in sympathy with the rights of Third World countries, traditional farmers and tribal peoples. Articles 3 and 15 recognise each country's sovereignty over its genetic and biological resources. In order to guard against biopiracy it requires that any person or corporation who wishes to gain access to these resources must obtain the consent of the host country (Article 15.5). This is good news for Third World countries that are rich in biological resources. It is not such good news for the pharmaceutical and agribusiness corporations that would like access to these resources free of charge. The Convention is particularly mindful of the role played by tribal people and traditional farmers in enhancing and maintaining biodiversity down through the centuries – Articles 8(j) and 15. It also affirms that the 'conservation of biological diversity is a common concern of humankind'.

Article 27. 3 (b) of TRIPs will effectively negate all these provisions and therefore CBD and the other milestones along the way to patenting, like PVPs and MTAs, should be opposed by every possible means.

It is worth noting that, while the US has pushed TRIPs in every possible forum, it has not yet signed the CBD. The US

Embassy in Thailand sent a strong letter to the Thai government when it began to draft legislation to protect its indigenous medical knowledge. The letter stated that the new legislation was in breach of the TRIPs Agreement. Many developing countries fear that if they do not bring in TRIPs-like legislation they may be put on the United States' Super 301 'Watch List' for free-trade offenders.

People who are campaigning against TRIPs ought to promote the Convention on Biological Diversity to ensure that there is a fair and equitable distribution of financial benefits derived from biological and genetic resources. This is the place to design mechanisms, including financial remuneration, which will reward individuals and companies for their investment and creativity in developing new products.

We also need to ensure that public research institutes protect the interests of poor, Third World farmers, and promote genuine sustainable agriculture. They can do this by protecting biodiversity and securing the rights of communities over their own biological resources and indigenous knowledge.

Given the huge impact of transnational corporations on global economic and political decisions, we need to develop effective international codes of conduct to monitor and regulate the activities of TNCs that control food and medicines. These codes must protect the rights, livelihoods and food security of the peoples of the world. Where they are breached there must be a mechanism to penalise TNCs in the courts.

Governments must ensure that food security, nationally and globally, does not escape from their control into the control of the corporate world. It is also essential that a public debate on the ethical issues involved in patenting take place before any international trade organisation attempts to promote a global patenting regime.

TRIPs FAVOUR THE RICH

Throughout this book, I have given many examples of how governments and big business have colluded to promote the vested interests of TNCs. TRIPs, as it now stands, is a perfect example of such collusion. It favours the rich and penalises the poor in the two most important aspect of life – food and medicine. Despite all the development rhetoric about eliminating poverty and banishing starvation, TRIPs institutionalise the economic dependence in which most poor countries are now trapped. According to the World Bank, the Uruguay Round of GATT did not benefit poor countries as the drop in tariff protection benefitted rich countries. Poor countries did not have armies of civil servant pouring over every text to see whether they would improve the situation for their economies. Many small countries had only one person each in Geneva to follow the complex negotiations. Very often, these Third World delegates were not fully consulted when important decisions were being taken. Most fair-minded people would agree that many countries didn't understand what they were signing up for when they accepted the TRIPs aspect of the Uruguay Round of GATT.[70]

Third World countries should be encouraged to walk away from TRIPs in its present form. The US, under President George W. Bush, had no problem walking away from the Kyoto Protocol on climate change because he considered it to be against US economic interests.

In any new agreement, priority must be given to protecting human health, poor subsistence farmers in the Third World, and the wider environment. One of the main planks in such a review would be the clear affirmation that the genetic code of all living beings ought to be considered the common property of humanity and the Earth and that no individual, state or company should be allowed to appropriate it.

---

70. Rana Foroohar, 'The Poor Speak Up' *Newsweek*, February 11, 2002, p. 32.

TRIPs are not only unjust; they is also hypocritical. While they fail to protect the genetic resources of the South, they facilitate the patenting of genetic resources that benefit giant Northern multinational corporations. This is not 'free-trade' as envisaged by Adam Smith. In fact it is a ploy to create global monopolies that Smith would abhor.

The WTO meeting at Doha in November 2001 was named the Development Round. It was nothing of the sort. Caroline Lucas, a Green MEP, wrote of it: 'developing countries were already furious before they arrived because the negotiating text drawn up in Geneva was weighted entirely in the interests of the rich North. But that was nothing compared to the ruthlessness of the negotiation tactics employed against them.' [71] Rich nations threatened to withhold official aid and debt-relief unless poor countries signed up to the new round of trade talks. The EU lobbied hard for the right to dump subsidised agricultural produce in poor countries – even though this has a devastating impact on local farmers. The industrial sector in poor countries was also hit and undermined. In Senegal, almost one third of manufacturing jobs have been lost because of pressure in a previous round to cut industrial tariffs by 50 percent.

The media presented the Doha meeting as a triumph for developing countries. They pointed to the fact that poor countries have secured the right to buy cheap generic drugs for the health needs of their poor. But even this turned out to be a Pyrrhic victory. While poor countries will be allowed to buy generic drugs by 2005, countries manufacturing them will be forbidden to sell them. Unfortunately, no action was taken to prohibit the patenting of life-forms in order that poor countries can protect local community rights to seeds from biopiracy by transnational corporations.

The WTO trade talks in Cancun in Mexico collapsed on September 15, 2003 without agreement being reached, mainly

71. *The Guardian*, November 21, 2001.

because of the inflexible and unreasonable demands made by both the EU and US. Rich countries wanted to complete the General Agreement on Trade and Services (GATS) especially regarding investment, competition and government procurement contracts. This would have facilitated transnational corporations moving into new areas of business in Third World countries, untrammeled by national legislation on sourcing a percentage of materials locally or on employing local people. It would increase business opportunities for Northern banks, construction, water, sewage, media and telecommunication companies to penetrate new lucrative markets. In return for this, rich countries were unwilling to give any substantial concessions to poor countries, especially in agriculture.

In the past, rich countries could bully poor countries into accepting unfavourable agreements by threatening to withhold foreign aid or deny them access to their domestic markets. This time, a group of twenty-one countries, led by China, India and Brazil, refused to be browbeaten into accepting the crumbs that the North had to offer. Other poorer countries from Africa, the Caribean and the Pacific also held out against US and EU pressure. Because of this deadlock, the talks collapsed without resolution. The collapse of multilateral trade talks is not a good thing in itself. Rich countries could adopt a divide-and-conquer approach and pursue bilateral trade deals with individual countries, thus undermining the multilateral approach where, at least in theory, the interests of small, poor, vulnerable countries could be protected. Still, no deal is better than a flawed one that would further exploit Third World people.

Our hope can be that, during the next two years of the Doha round, rich countries will begin to see that global poverty, malnutrition and environmental degradation will not be overcome unless rich countries assume their obligations to make sure that the Doha round of the WTO is truly focused on the needs of poor countries. Allowing US farmers to dump subsidised cotton in West Africa and thereby undermine the livelihood of ten million poor farmers cannot be considered a development initiative. Now that there is some unity among Third World countries and that they are now more and more aware of the implications of TRIPs Article 27.3 (b) one can expect that discussion around these issues will intensify before the next WTO ministerial meeting in two years time. [72]

The fact that life and death issues for tens of millions of people are affected by decisions made by ministers at a forum like the WTO was underlined by the suicide of the Korean farmer Lee Kyung-hae. Over the years he had turned a patch of harsh mountain into a thriving farm. His livelihood was totally undermined when cheap, foreign food was allowed into Korea. He lost his farm and was snowed under with debt. Lee believed that 'the multinational and big governments that control the WTO are pursuing a form of globalisation that is inhumane, farmer-killing and undemocratic. It should be stopped immediately, otherwise the false logic of neo-liberalism will perish the diversities of global agriculture and bring disaster to all human beings'. [73]

The umbrella organisation for Catholic development agencies, Cooperation Internationale par le Developpment et le Solidartité (CIDSE) has made proposals for changing the exploitative relationship between rich and poor nations, proposals that deal with many of the issues discussed in this book, from corporate power to patenting of life. CIDSE calls for:

1. Rewriting Article 27.3 (b) to exclude all life forms from patenting and removing the requirement for plant variety

---

72. Oisin Coghlan, 'Poorers States finally Call Time at Trade Talks', *The Irish Times*, September 16, 2003, p. 16.
73. Jonathan Watts, 'Fields of Tears', *The Guardian* (Supplement G2), September 16, 2003, p. 3.

protection with a moratorium on implementation in the interim.

2. (a) Ensuring that the provisions of the CBD take precedence over TRIPs Article 27. 3(b).

    (b) Ensuring that a revised TRIPs Agreement contributes to achieving internationally agreed development targets.

3. Introducing the principles of special and differential treatment for unequal parties contained in the WTO articles into the TRIPs agreement.

4. Greater equity and maximum transparency in the WTO negotiation process with adequate representation of all stakeholders in such negotiations.

5. The introduction of a strong anti-trust code into the WTO.[74]

---

74. *Biopatenting and Threat of Food Security*, CIDSE (International Cooperation for Development and Solidarity, Rue Slevin 16, B-1000 Brussels, Belgium, p. 27

Conclusion

THE FOCUS throughout this book has been on one of the most important and necessary realities in life – food. Humans have lived happily without mobile phones, cars or televisions; but we cannot live without food. That is why we must look carefully both at the current technologies involved in agriculture and at who controls them.

This book chronicles how gigantic agribusiness corporations have, during the past sixty years, increased their control of food production and, in the process, reaped enormous profits.

These corporations have promoted genetic engineered food, not because there was a demand for it from either consumers or farmers, but because it would give them greater control of agriculture and thus a guarantee of increasing profits. Unlike mobile phones, there is no market saturation for food. People need to eat every day.

Genetic engineering is still an imprecise technology, with the potential to cause harm to humans and the wider environment. For that reason, I have argued that the precautionary principle demands that there be no deliberate release of GE organisms until the technology is much safer. Testing mechanisms for this potentially dangerous technology must be taken out of the hands of the corporations who benefit financially from it, and returned to independent, well-funded government and international agencies who will promote the interests of the common good rather than the sectional interests of biotech corporations.

In Chapter Seven I insist that the patenting of living organisms – which has been promoted by corporations, and is now enshrined in Article 27. 3 (b) of GATT – is immoral and that it is another mechanism for exploiting the poor of the world.

I would like to see the Christian Churches and Christian development organisations taking a much stronger stand against patenting life. Food and life are at the heart of the Christian message. In the Eucharist we break bread and share it in the memory of the life, death and resurrection of Jesus. Our celebration of the Eucharist challenges us to work for a world where food is readily available for all and where it is produced in an environmentally sustainable way. Genetically engineered, patented food does not meet these criteria.

Further Information

ONE GOOD WAY to keep abreast of developments in this rapidly changing area is to check regularly with a number of websites.

The Union of Concerned Scientists is an alliance of leading scientists who are dedicated to promoting a healthier environment and a safer world: www.ucsusa.org/agriculture/biotech.html

Greenpeace International is a global environmental non-government organisation: www.greenpeace.org

Third World Net gives excellent coverage of North-South issues: www.twside.org.sg

The Edmonds Institute disseminates information about biotechnology: www.edmonds-institute.org

The Guardian: www.guardianunlimited.co.uk

GeneWatch UK: www.genewatch.org

Corporate Watch: www.corporatewatch.org.uk

Norfolk Genetic Information Network: http://ngin.tripod.com/

Friends of the Earth: www.foe.org.uk

The Ecologist: www.theecologist.org

Genetic Resources Action International: www.grain.org

Action Group on Erosion, Technology and Concentration: /www.www.etcgroup.org

Columban missionaries website: www.columban.com

VOICE: www.voice.buz.org

Oxfam: www.oxfam.org

Genetic Resources Action International: www.grain.org